An Integrated Model of Transport and Urban Evolution

Springer
Berlin
Heidelberg
New York
Barcelona
Hong Kong
London
Milan
Paris
Singapore
Tokyo

Wolfgang Weidlich · Günter Haag (Eds.)

An Integrated Model of Transport and Urban Evolution

With an Application to a Metropole of an Emerging Nation

With Contributions by: F. Englmann, K. Grützmann, G. Haag, P. Nijkamp, Y.S. Popkov, A. Reggiani, T. Sigg, W. Weidlich

With 44 Figures and 25 Tables

Springer

Prof. Dr. Dr. h.c. Wolfgang Weidlich
University of Stuttgart
Institute for Theoretical Physics
Pfaffenwaldring 57
D-70550 Stuttgart, Germany

Prof. Dr. Günter Haag
Steinbeis Transfer Center and University of Stuttgart
Applied Systems Analysis
Rotwiesenstr. 22
D-70599 Stuttgart, Germany

The publication of this book has been sponsored by DaimlerChrysler AG.

ISBN-13: 978-3-642-64317-0 e-ISBN-13: 978-3-642-60240-5
DOI: 10.1007/978-3-642-60240-5

Library of Congress Cataloging-in-Publication Data
Die Deutsche Bibliothek - CIP-Einheitsaufnahme
An integrated model of transport and urban evolution: with an application to a metropole of an emerging nation; with 25 tables / Wolfgang Weidlich, Günter Haag (eds.). With contributions by: F. Englmann ... - Berlin; Heidelberg; New York; Barcelona; Hong Kong; London; Milan; Paris; Singapore; Tokyo: Springer, 1999
ISBN-13: 978-3-642-64317-0

Hardcover-Design: Erich Kirchner, Heidelberg

SPIN 10734554 42/2202-5 4 3 2 1 0

Foreword

Gone are the days when mobility was nearly always a question of having a vehicle. Today the issue of road capacity is becoming ever more pressing. Even the safest, most comfortable and 100% emissions-free vehicle is only of limited use if it is stuck in a traffic jam. Mobility is a key human need and an important factor in the economy. It is a matter of logic that a company like DaimlerChrysler should make every endeavor to safeguard mobility, thereby fulfilling humanity‘s economic, social and environmental needs.

Nonetheless, traffic and mobility problems are the inevitable result of a concentration of people and markets. Bombay, Lagos, Shanghai, Jakarta, Sao Paulo, Cairo, Mexico City – virtually half of the world's population is urban-based, and the majority live in the metropolitan regions of the Third World. The mega-cities in the so-called developing nations are facing a dramatic increase in traffic levels. Gridlock looms on the horizon. Should traffic-choked streets become a permanent and daily occurrence, economic development will be held in check and pollution will spiral.

There can be no discussion as to the extremely high importance of these problems for society. DaimlerChrysler is certainly well-equipped to meet the challenges posed: As a global corporation, our products cover all transport areas. We have therefore seized the initiative by means of this study "An Integrated Model of Transport and Urban Evolution", which does not deal with purely commercial matters. Whilst emphasizing DaimlerChrysler's competence in the core area of mobility, the study is also the manifestation of a corporate philosophy in which worldwide social responsibility is accorded a central role.

Once more, here is one of the focal questions: How can the requirements for mobility be guaranteed and optimized to ensure sustained effectiveness? Scrupulous research into the role of traffic in metropolitan infrastructure development is the first step toward an answer.

I am convinced that one of the outstanding features of this study is its comprehensiveness. The traffic situation and the related factors were identified, recorded, sifted and reported in an expert manner. Dynamic mathematical models helped to ascertain the links and processes of interaction between traffic, demographic, economic, resource and environmental development factors. I might rightly add that we helped towards the success of the venture by bringing our wide-ranging knowledge and competence in traffic research and development to the shared project. Prior to this, DaimlerChrysler had already been intensively engaged in designing traffic concepts for urban clusters. The experience we had gained from very similar ventures in Berlin, Athens, Bangkok, Hanoi and Cape Town stood us in good stead.

The current study supplied us with fresh realizations and much food for thought, all of which were invaluable. Nanjing has yielded decisive progress – we now know for certain that these results are of tremendous relevance for places far beyond the South Chinese metropolis. In theory, the same procedure may be applied to other cities.

Having stressed the purely scientific success of this study, I should like to point up another key aspect, namely the successful combination of science and economics. The study has proven something of a textbook example for science sponsorship – and even the sponsorship itself reaps the benefits. This instrument is becoming increasingly attractive.

There is another feat which should not be overlooked, namely the trouble-free coordination of scientists from various specialists fields, from different divisions of a large corporation, and operating from locations in Russia, Belgium, The Netherlands, Italy, China and Germany. The following pages will show just how productive this mixture of expertise really was.

I would like to take this opportunity to once again thank all those who made the study such a success. I should like to mention first Prof. Weidlich from the Institute for Theoretical Physics at the University of Stuttgart and Prof. Haag from the Steinbeis Transfer Center for Applied Systems Analysis for their exemplary dedication in coordinating and implementing the project. Special thanks are also due to Nobel Prize winner Prof. Prigogine in Brussels and to Prof. Haken from the University of Stuttgart. Furthermore I would like to thank the Professors Nijkamp from the Free University of Amsterdam, Reggiani from the University of Bologna, Englmann from the University of Stuttgart and Popkov from the Academy of Science of Russia. Last but not least, special thanks go to Prof. Wang Wei and Prof. Deng Wei from the Southern University of Nanjing for their cooperation in the project and provisions of indispensable data.

Dr. Uli Kostenbader, DaimlerChrysler AG

Preface

It is a privilege to write a short preface at the occasion of the publication of the project *"An Integrated Model of Transport and Urban Evolution"* due to Professor W. Weidlich and his co-workers. It gives me the occasion to come back to problems which the late R. Herman and I have studied decades ago. Of course time is going and this project contains many new aspects.

The importance of the problems studied in this book is obvious. The interaction between transport and the spatial activity systems is critical in shaping urban and regional economic and social development. Understanding such interactions and the associated evolutionary patterns should guide infrastructure investment decisions and the development of policies regulating the operation and use of the transport systems, as well as measures that affect the amount and nature of spatial activity. Such actions can have profound economic, social and environmental consequences, especially in rapidly developing contexts such as the Nanjing urban region in China. This context is highly dynamic, and the land uses transport system may evolve in dramatically different ways depending on exogenous as well as endogenous forces under different policy actions.

The difficulties to treat such problems are enormous. I think that the authors would agree that much remains to be done. However, the project presented here is a promising starting point and the results already obtained allow us to conclude that additional progress may be achieved in the near future.

Stuttgart, July 1999 Prof. Dr. Dr. h.c. mult. Ilya Prigogine

Preface

This book summarizes the results of a study that was sponsored by the Daimler-Benz AG (now DaimlerChrysler AG) and that was performed under the guidance of Prof. W. Weidlich, University of Stuttgart, and Prof. Haag, Steinbeis-Transferzentrum Angewandte Systemanalyse, Stuttgart. The aim of this project is to provide emergent nations with scientific support on urgent problems in traffic and city development.

The project presented here is largely based on theoretical concepts developed by Haag and Weidlich in Stuttgart, whereby the international cooperation with further centers proved very helpful. The city of Nanjing, the former capital of China and an important traffic junction, was chosen as the subject of study, whereby both the data provided from the Chinese partners and discussions with them proved to be important. The researchers involved in this project must be congratulated for the excellent job they did.

The study presents several scenarios on traffic and urban development in Nanjing. As a result of these scenarios, some concrete recommendations for the further development of Nanjing are given, and I am sure that these will be of great help for the planners and decision makers in China. Thereby, it must not be forgotten that because of the very many economic, social, ecological, and other aspects, such decisions are, eventually, political.

Stuttgart, July 1999 Prof. Dr. Dr. h.c. mult. Hermann Haken

Contents

Part I

Introduction into the Economic and Traffic Situation of China in General and of Nanjing in Particular

1. Introduction

October, 1st 1996 a new kind of co-operation between the university of Stuttgart and industry was initiated for the first time: The DaimlerChrysler AG assigned a scientific sponsoring-project with the subject **Traffic and City Development in Emergent Nations** to the University of Stuttgart.

"The Third World faces a big challenge with respect to the development of the transportation of persons and goods, especially in the very fast growing conurbation areas. Currently there exists a good chance for the emergent nations to create well-designed transportation and urban/regional structures concerning their technology, economy and ecology."

This statement by Prof. Rolf Scharwächter, the general representative of the DaimlerChrysler AG explains the involvement of such a globally active enterprise in support of the ITEM project whose task is the development of forecasting future traffic scenarios taking as a model the Southern Chinese metropole of Nanjing City. The project is coordinated by Prof. Wolfgang Weidlich of the II. Institute of Theoretical Physics of the University of Stuttgart in cooperation with Prof. Günter Haag of the Steinbeis Transfer Centre Applied System Analysis STASA of Stuttgart utilizing the expertise of economists, regional scientists and traffic experts from Stuttgart, Bologna, Amsterdam, Moscow and of course Nanjing.

1.1 The Aims of ITEM (Integrated Transport and Evolution Model)

The primary aim of the project is to understand the essential characteristics of the interrelated dynamics of the traffic system and the regional settlement structure of metropoles in emergent nations in order to be able to make scenario-based forecasts.

It is the intention of the project to yield a threefold benefit:
First of all we hope that the results of the project will be useful for the examined region, i.e. the city of Nanjing. The calculations, simulations and forecasts will of course be made available to the Faculty of Traffic and Transportation of the South East University of Nanjing which has provided us with the empirical data. In this way the region of Nanjing receives in-

formation about the traffic situation to be expected under different planning alternatives.

Secondly, the project is useful for applied research, because it gives answers concerning the flexibility and quality of the modelling procedure in dependence upon the available empirical data.

Thirdly we hope that the project will also be useful for the sponsor, the DaimlerChrysler AG, because its results will give insights into the development of a region belonging to an important market of growing importance.

In order to fulfil these high expectations the dynamic model to be used must be very flexible and applicable not only to the analysis of the current situation but also to forecasts via scenario techniques, even if not all desirable data are available.

The city of Nanjing proves to be an excellent example because of the following aspects:

Nanjing has about 2.6 million inhabitants in the city and more than 5 million within the region of Greater Nanjing. It is therefore not too large but as a traffic junction between Beijing in the North, Shanghai in the West and Hongkong in the South, it plays a very important role within the national transportation system. Nanjing is located on the Yangtse River, ca 300 km to the West of Shanghai; it has the largest overseas harbour in China and owns one of the few bridges over the wide river. In the South of the city a large international airport is currently under construction. Moreover, Nanjing is an administrative centre as well as a centre of science. The Southeast University of Nanjing is one of the five most important universities of China.

The rapid economic evolution in Nanjing causes large Chinese as well as foreign investment. This can be seen in the CBD (central business district) where many new buildings and skyscrapers are currently erected. The creation of new working places simultaneously induces a displacement of buildings with less floor capacity. It is obvious that this course of events will produce increasingly unmanageable problems unless a complementary development of the transportation system.

The ITEM project intends to become an instrument for the analysis and forecasting of traffic and urban evolution and urban/regional planning.

In particular, the model should be able to give answers to the following questions:

- In which way does the relationship between different traffic modes change?
- Which consequences does the separation of landuse induce?
- What will happen when the availability of certain traffic modes (car, bike, ...) changes?
- What are the consequences of the erection of a second bridge over the Yangtse River?

1.2 Description of the Project

In the following a short description of the modelling concept used in project ITEM, developed at the Institute of Theoretical Physics of the University of Stuttgart in co-operation with the Steinbeis Transfer Center Applied System Analysis (STASA), will be given.

In order to understand the complex processes and interactions between traffic and urban development, the partial systems are analysed firstly by stochastic decision models. For that purpose the “area under consideration” is divided into homogenous regions (cells or locations). The development of these regions is described by "macro-variables", the so-called regional configuration. Examples of these variables are all those which describe the traffic infrastructure in the cells but also the current distribution of the population over the single cells, the distribution of lodgings, factories, shopping centres, leisure centres, etc.

The modelling concept establishes a connection between the dynamics of these macro-variables and the behaviour of the individuals of the considered population ("microlevel"). To describe the decision behaviour of actors the following two factor sets are essential: on the one hand the attractivity differences of the single regions as "driving forces" of the decision processes; on the other hand a global flexibility of the actors as well as distance effects between the cells.

The attractivity of a single cell is determined by its characteristics such as population number, number of working places or purchase possibilities. The distance effects are influenced by the geographical distance as well as by the "social" distance between the single cells. That means for instance that a household is better informed about shopping possibilities in the neighbourhood than in cells which lie far away.

Generally, the interactions between micro- and macrolevel are captured as follows:
On the one hand the dynamics on the macrolevel - i.e. the development of the traffic and of the regional settlement structure - is determined by the behaviour of the individuals on the microlevel. On the other hand the attractivity differences between the cells which depend on the macro-variables influence the decisions of the individuals.

Apart from rational motives of the actors several elements of uncertainty, e.g. irrational behaviour as a result of insufficient information, have to be taken into account. Hence, the description of decision processes is based on a simultaneously stochastic and dynamic decision model, the so-called master equation approach. From the master equation there can be derived meanvalue-equations which are the basis of the present ITEM model.

The traffic system as well as the settlement structure and the population of Greater Nanjing form a complex intertwined system. Its dynamics take place on different time scales but are modelled making use of the same principles:

a) The daily flows of traffic in Nanjing are the result of very fast decision processes of the actors in realising a trip between two traffic cells (origin - destination) with a special purpose. Decision processes for a certain destination, the moment of the setting out, the mode of transportation, the choice of the route etc. take place on a very short time scale. In the model the different traffic flows with respect to the different trip purposes are related to the different modes of traffic. The model includes the attractivity differences between the cells caused by different characteristics of the cells (e.g. population, workplaces, factories, lodgings, etc.) as well as distance effects or "resistance effects" expressed by different travelling times (implying different costs) between the cells.

b) On the other hand the development of the regional population is a process on a long-term time scale. The population distribution changes because of migration, i.e. moves between the cells. The equations of motion which describe the migratory behaviour contain transition rates, i.e. migration flows between the cells. These flows depend on distance effects and attractivity differences as a result of different regional advantages (e.g. rents, available living space, traffic links etc.).

The corresponding development of the settlement structure is also a very slow process. The attractivity differences for the evolution process of settlements are influenced by economical factors such as rents, costs per m^2 or position factors (being within reach, etc.).

The available empirical data such as traffic flows between the regions, population numbers or migration flows are based on the traffic system as well as on the development of the population distribution. The system parameters as attractivities and distance (or resistance) parameters are then determined with the help of a non-linear estimation procedure. In a further step the attractivities can be connected with the macro-variables or key-variables making use of a multiple regression. At the same time the relevance of the single key variables is determined. The last two steps consist in numerical simulations and recommendations derived from these simulations.

The report will present both the concrete study of the development of the regional transportation system for the Region of Greater Nanjing and the general structure of the modelling procedure. The presentation is organised as follows:

Part I (Introduction into the economic and traffic situation of China in general and of Nanjing in particular):
Section 2 gives an overview of the regional development of China. It offers some background thoughts on recent changes in China, seen from a spatial and transportation viewpoint. Section 3 deals with the economic and geographical situation of the Region of Nanjing. Specifically, it focuses on the current housing situation and the industrial aspects of the City. In section 4 the traffic development in China, especially within the region of Nanjing is presented. In addition trends and scenarios in the transportation sector can be found there.

Part II (Description of the model and of the adaptation to the available data):
The sections 5-7 form Part II of the report. Section 5 describes the ITEM model, the transportation and the urban/regional levels and the method of estimating the parameters. In section 6 the data of the project which were available to us are listed and explained, whereas section 7 describes the

adaptation of ITEM to this specific data situation in Nanjing. At the end of the latter section a summary of the relevant algorithms is given.

Part III (Presentation of the scenarios, results of the calculations, conclusions and recommendations):
Part III contains in section 8 the methodology, i.e. the analysis and forecasting of the traffic and urban/regional situation of Nanjing. Section 9 presents the results in the two sectors. In section 10 the scenarios are described, analytic and numeric results of the scenario technique are given and the results are discussed. Section 11 contains conclusions which can be extrapolated from the results as well as our general recommendations to Nanjing.

Appendices (Detailed mathematical description of the model, Figures of the results):
The appendix A1 exhibits the ITEM model in mathematical detail. Appendix A2 shows the maps of Nanjing in different scales with respect to different locations. Finally, in the appendices A3 and A4 the results of the scenario calculations are elucidated.

2. China in a State of Flux

Peter Nijkamp, Free University Amsterdam
and Aura Reggiani, University of Bologna

Abstract

This paper offers some background thoughts on recent changes in China, seen from a spatial and transportation viewpoint. It calls attention to a great many uncertainties involved in depicting the future of the Chinese spatial economy and argues that scenario analysis is a meaningful research strategy for coping with an uncertain future. The paper then sets out to explain a systematic approach to scenario analysis and the design of future images. The paper concludes with a multicriteria analysis of various possible spatial development trajectories in China.

2.1 The Changing Scene

Our world is increasingly moving towards an urban world and an urban life style. The Mega-Cities Project description (1991) states: "By the year 2000, more than half of the world's population will live in cities. It is projected that 23 of these cities will be "Mega-cities" with more than 10 million people each. Despite their varying political, economic, social and cultural characteristics, all of them face a common problem: they must provide workable cities for unprecedented numbers of citizens within limited budgets and severe environmental constraints. The time is right for new approaches."

One of the countries facing a rapid urbanization process is China. In a recent article, Bradbury and Kirkby (1996) make the following observations: "The scale of China's population - currently standing at around 1.2 billion - is well known, as are the nation's fertility restriction measures. With a significant relaxation of the one-child policy in rural areas during the late 1980s, China faces a further substantial increase in population, to an estimated 1.3 billion by 2000, and to 1.5-1.6 billion by 2050 (State Planning Commission, 1994). For all Chinese commentators these statistics dominate every consideration of the global environment."

The development of township and village enterprises is a crucial facet of contemporary Chinese urbanization, providing the basis for the expansion of tens of thousands of small towns and villages with new enterprises, infrastructure and housing. However, the present process of change has also involved a significant growth of major cities. The most recent 15 year interval has seen a rural-to-urban population shift equal in scale to that of the 1950s. Now almost 30 percent (350 million) of the population is officially, and plausibly, classified as urban compared with just 18 percent (172 million) in 1978. Throughout the urbanization spectrum similar environmental concerns emerge, including the loss and degradation of agricultural land, inadequate resource supply, poor waste disposal and low infrastructure provision. The population-equation, a recurrent theme throughout recorded history in China, remains a significant element in the debate about environment and development.

Atmospheric pollution, which the document acknowledges to be one of the most serious environmental problems in urban areas, merits a chapter to itself. As a signatory of the UN Framework Convention on Climatic Change and the Montreal Protocol on Ozone Depleting Substances, China now has international obligations in this area. The single most important source of atmospheric emissions is coal, a generally low-quality fuel which is used inefficiently and with little attention to pollution abatement. Motor vehicles are a supplementary but increasingly important source of urban air pollution.

It goes without saying that the current vibrant development of China - caused by its open door policy - leads to a series of hitherto unprecedented phenomena, such as a rise in car mobility, and large-scale commuting as a result of the spatial divergence between urbanization and industrialization. In a study on the free trade zone (FTZ) of Shanghai (another manifestation of the rapid changes in China), the authors (see Massey et al. 1996) note: "On the face of it, at the current relatively early stage of deve-lopment of the transport infrastructure of the Shanghai area, the Shanghai Municipal government faces a choice between either drawing upon the recent experience of more developed countries or ignoring that experience and ultimately learning the same lessons in the longer term as a consequence of ma-king the same mistakes that governments in more developed economies have made in the past and continue to make today. In simple terms, there appears to be a real danger that the pressures for an increase in car ownership and use will go unchallenged. This is reflected in what can be viewed as the 'access problem' faced in the development of the FTZ a problem which threatens to become acute in the near future".

The rise in geographical mobility and the related transportation problems are a direct consequence of open door policy. "To get rich is glorious", Deng Xiaoping's slogan defined the direction of the new development strategy in China after 1978. The resulting high growth rates of the Chinese GNP of about 10% per year have been noticed around the world. A comparable speed of economic development has been realized only in the newly industrialized countries in South-East Asia. Many people in China have benefited from reforms and the living standard of these people has been improved in many respects. The creation of free markets for agricultural and industrial products, the acceptance of private entre-preneurship and many other reform steps have created strong and effective incentives for individuals. The decentralization of the decision making power has given the provinces the chance to take more responsibility for their own economic development" (Klotz and Knoth 1997).

A major problem in the current Chinese development towards rapid modernization is the lack of an appropriately sophisticated infrastructure. This is even more severe, since the gradual shift towards a market form causes drastic shifts in economic activities (see Kim and Knaap 1997). Although there are many plans, congestion due to the variety of modes is increasingly becoming a problem. Environmental pollution caused by the rise in motorized traffic is another cause of high social costs. In the sequel of this paper we will focus in particular on the transportation sector.

2.2 Unknown Pathways

Transportation is manifesting itself in a force field of conflicting objectives, such as economic efficiency, spatial-economic development, environmental protection and safety. In this context, the concept of sustainable transport or sustainable mobility has been introduced in recent years (see for a full description Nijkamp et al. 1997).

In light of the great many uncertainties surrounding future transport systems, it has become customary to employ scenarios as a vehicle for a structured communication process with policy-makers. Scenarios have gained much popularity in the past decade, as they are richer in scope than conventional forecasting tools and offer a more 'open world' to policy-makers. A scenario may be defined as: "a tool that describes pictures of the future world within a specific framework and under specified assumptions. The scenario approach includes the description of at least two or more scenarios designed to compare and examine alternative futures" (CEC, 1993).

In a recent paper on images of future transport in Europe, Drebord et al. (1997) state: "Scenario methodologies that do not concentrate on the desirable scenarios, may be labelled *explorative*. Such methodologies are reasonable when essential parts of the system under study cannot be controlled by policy measures. The scenario analysis may then help in the development of a strategy of flexibility and adaptability, i.e. how to cope with uncertainty and surprise.

On the other hand, in a planning context where the actors involved may greatly affect the development, a reasonable strategy would be to attempt to shape the future according to what is preferred, instead of just adapting to what may emerge. Backcasting studies will then be of interest.

Turning to the transport sector, it is obvious that transport policies may have a great impact on the development of transport in the long run, but it is also true that driving forces and conditions beyond the control of transport policy makers will have an impact as well, and will also influence the conditions for policy making. This is the question of *external* vs. *internal factors*, which is highly relevant in the context of transport policy analysis".

Factors which are external to the transport sector are **inter alia** institutional/political developments inside (or outside) the country at hand (e.g., drastic policy shifts), or the introduction of new technologies. Clearly, as soon as such new technologies are developed by the transport sector itself, they become internal to the structure of that sector. Examples are improved vehicle technology or alternative fuels (e.g., electric cars, methanol).

In the post-war period we have witnessed a steady increase of the action radius of mankind, not only for daily home-to-work trips, but also for business trips or leisure trips. In the age of globalization it seems likely that a continued pressure will exist towards higher mobility levels.

In this context, it ought to be emphasized that physical movements as such are not a bad thing. Of course, they are costly, but they also create many economic benefits to both consumers and business life. The problem is however, that there are many social costs involved in the form of pollution, accidents etc. which are not (fully) charged to the source of the externalities. Thus, the main question for many governments nowadays is whether it will be possible to ensure that mobility rates are compatible with environmental sustainability criteria. This would mean that the so-called **decoupling** hypothesis would have to be implemented, which implies a delinking of economic growth and the environmental decay caused by the transport sector. This would of course require drastic changes in our modes of production, consumption, transportation and technological innovation. Clearly, there are many uncertainties involved in developing such new (policy) strategies, and hence it seems wise to rely on appropriate scenarios which might depict some of the future spatial images of our world and which would also allow us to identify bottlenecks in the developments to come.

In the Chinese context, it is clear that decisive external factors for transport scenarios are also formed by the demography, the economic policy (e.g., the open door policy), as well as by the settlement and industrialization policy. Thus, there are many uncertain factors which may impact upon the transportation system in China, not only at the national level but also at the urban level.

2.3 Some Evidence on the Chinese Transportation System

China is a vast country and it is thus no surprise that a significant part of the public budget is spent on transportation infrastructure. The country used to have only a relatively sparse density of highways and railways, but the situation has improved very much in recent years. According to some experts, China will have to multiply its transportation infrastructure investments by a factor of 20 in the year 2025, in order to progress with economic growth expectations.

Furthermore, the Chinese population will most likely increase by more than 30 percent by the year 2025, so that the total population will amount to approximately 1.5 billion people. Clearly, also a shift from rural to urban areas may be expected. There will be mass urbanization processes in the future. For example, it is expected that in one decade China will have three large metropolises with a population over 10 million each, viz. Shanghai, Beijing and Tianjin.

In addition, the new economic policy - thanks to its open door character - has been very successful and is expected to generate at least 7 percent growth annually. This will of course create much more mobility and consume much more energy. For example, energy consumption in China has in 1989 created 6.0 million metric tonnes of CO_2, so that serious environmental damage and climate change may be expected. This puts the notion of sustainable transport and sustainable mobility in question, as far as the Chinese contribution to global environmental change is concerned.

In a recent study, Huang and Chen (1994) argue that the present inadequate transport systems are becoming bottlenecks in various economic activities in China. They argue: "Since 1949 there has been great progress in transport systems in the People's Republic of China. The comprehensive transportation system which mainly includes five divisions, i.e. railway, roadway, waterway (including ocean shipping), civil aviation and pipeline, has basically been established. However, China's economic development is still being restrained by its present transportation conditions. The economic open policies have been generating rapid growth in floating populations and economic expansion which are accompanied by growing demand for transport of freight and passengers. As a result, the inadequate transport systems are becoming bottlenecks in various economic activities (for example, the movement of coal from mine to user, the transport of agricultural and light industrial products from rural to urban areas and the delivery of imports and exports).

The reason for this unfavourable situation can be divided in two. Firstly the excessive population, vast territory, unreasonable industry location and weak investment intensity on transport infrastructures. Secondly the wrong decisions involved in developing five transport modes. In the past 40 years, too much attention and corresponding investment were given to rail, and the superiority of the other four modes was ignored".

Some figures may be helpful to illustrate the vast problems China's transportation system is faced with (see Huang and Chen 1994).

Table 2.1: Length of Transportation Routes (1000 km)

Year	Railways	Roads	Waterways	Aviation Lines (intn.)	Pipelines
1978	48.6	890.2	136.0	148.9 (55.3)	8.3
1988	52.8	990.6	109.4	373.8 (128.3)	14.3
1990	53.4	1028.3	109.2	506.8 (166.4)	15.9
1992	53.6	1056.7	836.6	836.6 (303.0)	15.9
1993	53.8	1083.5	110.2	960.8 (2787.7)	16.4

Table 2.2: Passenger Traffic (million passenger)

Year	Railway	Road	Waterway	Aviation	Total
1978	814.91	1492.29	230.42	2.309	2539.93
1988	1226.45	6504.73	350.32	14.420	8095.92
1990	957.12	6480.85	272.25	16.590	7726.82
1992	996.93	7317.74	265.02	28.860	8608.55
1994	1087.38	9539.40	261.65	40.380	10928.83
1995	1027.45	10408.10	239.24	51.170	11725.96

Table 2.3: Passenger Traffic Volume (billion passenger-km)

Year	Railway	Road	Waterway	Aviation	Total
1978	109.32	52.13	10.06	2.79	174.30
1988	326.03	252.82	20.39	21.70	620.70
1990	261.26	262.03	16.49	23.05	562.80
1992	315.20	319.30	19.80	40.60	694.40
1994	363.60	422.00	18.40	55.20	859.10
1995	354.60	460.30	17.20	68.10	900.20

Table 2.4: Freight Traffic (million ton)

Year	Railway	Road	Water-way	Aviation	Pipe-line	Total
1978	1074.92	851.82	432.92	0.064	103.47	2489.46
1988	1449.48	7323.15	892.81	0.328	156.18	9821.95
1990	1506.81	7240.40	800.94	0.370	157.50	9706.02
1992	1576.27	7809.41	924.90	0.575	147.83	10458.99
1994	1630.93	8949.14	1070.91	0.829	150.92	11802.73
1995	1658.54	9403.87	1131.94	1.011	152.74	12348.10

Table 2.5: Freight Traffic Volume (billion ton-km)

Year	Railway	Road	Water-way	Aviation	Pipe-line	Total
1978	534.52	27.41	377.916	0.097	43.00	982.90
1988	987.76	322.04	1007.040	0.730	65.00	2382.50
1990	1062.24	335.81	1159.190	0.820	62.70	2620.70
1992	1157.60	375.50	1325.60	1.340	71.70	2921.80
1994	1245.80	448.60	1568.70	1.860	61.20	3326.10
1995	1287.00	469.50	1755.20	2.230	59.00	3573.00

Table 2.6: Average Distance of Freight Traffic (km)

Year	Railway	Road	Waterway	Aviation	Pipeline
1978	485	32	873	1521	416
1988	681	44	1128	2226	416
1990	705	46	1447	2218	398
1992	758	46*	1554*	2234*	399*

"In 1990, the number of civil motor vehicles has reached 5.514 million compared with 1.358 million in 1978 (816.2 thousand private vehicles). In 1991, the figures were 6.06 million and 960.4 thousand respectively.

In addition, the cargoes handled at principal seaports were 198.34, 490.25, 483.21 and 532.4 million tons in 1878, 1989, 1990 and 1991 respectively. In 1991, there were 1211 berths at principal seaports (296 with the capacity of handling over ten thousand-tons ships). The number of civil aviation aircrafts was 410, 413, 421 and 438 in 1988, 1989, 1990 and 1991 respectively. The number of international civil aviation routes was 40, 44, 44 and 49 in 1998, 1989, 1990 and 1991 respectively; meanwhile, the domestic routes were 310, 334, 393 and 403 respectively".

2.4 Domestic Scenarios for China

The domestic problems faced by the transport sector in China are twofold:

- how to cope with limited infrastructure capacity in a rapidly expanding economy in which mobili-ty of people and goods may show a significant increase in the future?
- how to cope with the problem of environmental sustainability in a rapidly car dominated country?

As a frame of reference, it may be interesting to report some objectives laid down recently in the Common Transport Policy of the EU:

- free movement of goods and persons
- development of a coherent, integrated transport system using the best available technology
- reduction of interregional disparities, inter alia by infrastructure construction
- achievement of sustainable patterns of development by taking care of the environment
- stimulation of higher safety in traffic
- encouraging social cohesion in the EU
- development of appropriate relations with third countries

A transfer of such objectives to the Chinese context would create formidable problems, as many of these objectives are mutually inconsistent and not compatible with the current economic conditions of China. In general, it seems plausible to make a distinction between three major distinct policy orientations which tend to be present in each modern policy. These are:

(i) Maximization of economic benefits through an efficiently operating transport system

(ii) Consideration of spatial equity motives through a reduction of interregional welfare differences through the provision of access to transport networks

(iii) Achievement of environmental sustainability and security through pricing mechanisms and proper regulations

These three major orientations for transport policy can be mapped out in a policy triangle (see Figure 1)

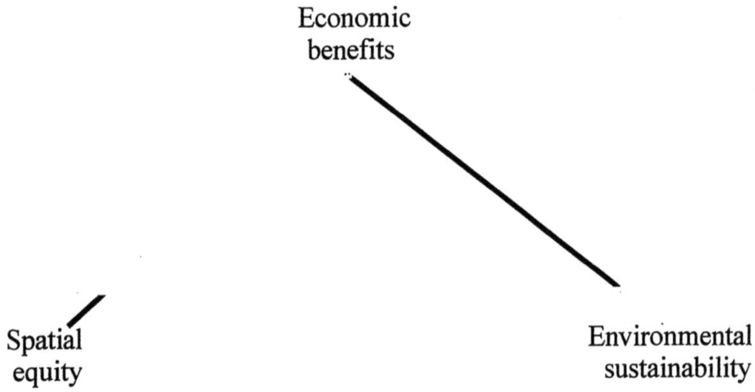

Figure 1: The triangular policy concern

We will now briefly describe each of these three motives for transport policy.

Economic benefits

The economic benefits of the transport sector (and hence of mobility of persons and goods) can theoretically be traced via its contribution to GDP per capita by reference to the fulfilment of Pareto - optimality criteria. In applied research, it is rather common to make use of cost-benefit analysis. In reality, several problems may be mentioned which are related to the specific nature of transportation. Transport is essentially an input (or a production factor), whose optimal allocation has to be determined together with other inputs such as labour, capital or energy (depending on its relative price). Transport as a distinct industry is also a contribution to economic growth and is in most countries responsible for some 6 to 8 percent of GDP. And finally, transport has often a derived demand which does not only have a productive, but also a consumptive character.

Spatial equity

Reduction of interregional disparities (or stimulation of socio-economic cohesion) is another major policy motive. For example, in the EU the main part of the Structural Funds is spent in less favoured regions. It turns out that transport infrastructure has normally a network nature, through which also less central areas can be linked to other areas. Consequently, spatial accessibility is a major concern in a balanced transport policy which serves the needs of less centrally located regions or cities. In some cases, it may be meaningful to define an accessibility performance index as the quotient of intra-regional accessibility over interregional accessibility.

Environmental sustainability

The environmental implications of the transport sector are rather severe. The externalities manifest themselves in the form of air pollution, toxic materials, noise, visual intrusion and solid waste. It has become common to make a distinction between strong and weak sustainability. Strong sustainability means that everywhere environmental standards are to be respected without allowing for substitution between various pollutants. Concepts like critical loads, maximum carrying capacity or critical threshold values apply in this case. Alternatively, weak sustainability allows for substitution by a relaxation of some pollutant emissions provided the overall outcome is still more favourable. Examples of policy strategies here are: no regret policy, double dividend, and maximization of maximum consumption per head over an infinite time horizon (the Solow criterion).

The policy orientations discussed briefly above may now be combined with a set of exogenously given (or assumed) developments for China. In the framework of our analysis, we will present two major scenarios which may be seen as decisive for the future spatial development of China. These two extreme framework developments are: **polarization** versus **co-ordination** on the domestic Chinese market. In Table 2.7 the various main features of these two frame scenarios are sketched (see also Hey et al. 1997).

Table 2.7: Two extreme frame scenarios for China

Polarization	Co-ordination
Institutional/economic	**Institutional/economic**
□ spatial desintegration	□ multiregional coordination
□ uncoordinated transport policy	□ balanced transport policy
□ lack of intermodal transport	□ stimulation of intermodality
□ uncertain growth perspectives	□ steady growth perspectives
Social	**Social**
□ lack of environmental care	□ environmental-benign transport
□ neglect of equity goals	□ interest in spatial cohesion

It is now possible to combine these two frame scenarios with the three policy orientations described, above so that a compound system of six images may be created. These compound images are depicted and described in Table 2.8; these are essentially a mix of frame scenarios and policy motives. Each of these images has been named after the main constituent elements. Their description follows a plausible reasoning based on likely spatial-economic and transportation impacts.

Table 2.8: Compound images of transport developments in China based on two frame scenarios

Competitive Regions (CR)	**Concerted Nation (CN)**
Economic benefit and polarization	Economic benefit and co-ordination
□ decentralization □ market orientation □ private investments □ private transport initiatives □ low interest in public transport □ unsustainable mobility	□ concerted transport policy □ market orientation □ private/public partnerships □ partly regulated transport market □ role for collective transport □ low environmental interest
Balanced Regions (BF)	**Harmonious Nation (HN)**
Spatial equity and polarization	Spatial equity and co-ordination
□ decentralization □ low market orientation □ limited privatization □ decline in public transport □ inefficient transport network	□ centralized transport policy □ no market incentives □ strong role of government □ interest in public transport □ spatially connected networks
Environmental Regions (ER)	**Sustainable Nation (SN)**
Environmental sustainability and polarization	Environmental sustainability and co-ordination
□ modest road investments □ emphasis on core areas □ completion HST network □ internalization of social costs □ low mobility growth	□ low road investments □ investments in new fuels □ focus on public transport □ road pricing □ stand-still for mobility growth

Based on the qualitative description given in Table 2.8, we may now design a qualitative 'virtual' impact table, where for each of the six images the expected (likely and plausible) consequences are depicted, based on expert opinion.

Table 2.9: Qualitative assessment table for various spatial images of China

	Economic benefits		Spatial equity		Environmental sustainability	
Types of spatial images	Inter-nalized Social costs	Decopling Transport-Growth	Increase Disparity	Unem-ployment periphery	Reduc-tion CO_2	Reduc-tion NO_x
CR	5	3	1	2	3	4
BR	3	2	4	4	2	2
ER	1	2	1	1	5	5
CN	5	4	2	2	5	4
HN	1	1	5	5	1	1
SN	3	3	1	3	5	5

1 = drastic worsening
2 = moderate worsening
3 = ambiguous result
4 = improvement
5 = drastic improvement

The information contained in Table 2.9 lends itself to a fruitful application of multicriteria analysis. In view of the ordinal rankings, in particular the so-called **regime analysis** seems to be suitable in this framework (see for full details Nijkamp et al. 1993). The application of the multicriteria regime method presupposes the availability of weights to be attached to each of the criteria given in Table 2.9. In the absence of such information we will undertake here some sensitivity experiments (see Table 2.10).

Table 2.10: Results (in terms of success scores) of priority sets for criteria on spatial images of China

	various priorities attached to policy criteria			
Spatial images	Unknown Priorities	equal weights	equity and environment	economic benefit
CR	.52	.60	.50	.60
BR	.42	.40	.30	.29
ER	.24	.20	.40	.30
CN	.92	1.00	.89	1.00
HN	.12	.00	.00	.00
SN	.77	.80	.90	.80

The results from this multicriteria experiment are highly interesting. It turns out that a **concerted national** policy for China has in most cases the highest priority, followed immediately by a **sustainable national** strategy. In all cases, a harmonious national strategy focused on spatial equity in the country is clearly inferior. The other images assume an intermediate position, in which the concept of competitive regions appears to assume a robust third position.

These scenario-experiments and related images can be used to undertake analytical experiments on possible policies for transport systems in China.

2.5 Political-Economic Scenarios for the Development of China

In China, successful reform policies, especially the introduction of "special economic zones", have resulted in a doubling of its per capita income over the past decade. A key characteristic of the strategy was extensive government planning aimed at the rapid substitution of foreign imports for domestic production. For this purpose, industry was heavily protected against foreign competition. According to a report of the World Bank, China may become in about 25 years the strongest economy of the world, if this country still develops at the current rate. The double-digit growth rates achieved in China in 1992 and 1993 raised concerns about the economy becoming overheated and inflation running out of control. The rate of inflation accelerated to 25% in 1994. The inflation moderated somewhat, following the introduction of China's austerity measures. A depreciation of the Chinese currency reduced the costs of imports from China. This inflation moderation was also supported by low international prices and costs of raw material imports. The estimated 13% growth in GDP in China in 1993 was buttressed by a rapid growth in industrial production. Growth was sluggish however, in the energy and transport sectors, raising concerns that their inadequacies could give rise to bottlenecks to further economic expansion. The rate of import growth exceeded that of exports. Almost 25% of Chinese exports went to Hong Kong, while exports to Japan accounted for 17%. In 1993, export growth slowed down for a number of reasons. These included high domestic demand and diversion of some of the export items for domestic use. The key economic issue facing China's leadership is how to liberalise the economy further and end the current inflationary spiral. Overheating is causing shortages of raw materials and infrastructure problems. Although the economy is liberalised, the political environment remains tightly controlled. The aftermath of Deng Xiaoping's death will lead to attempts to consolidate the leadership's grip on the economy. (see also Nijkamp and Vermond 1996).

Based on the above sketch of background developments one may envisage several future images for China, based on political-economic perspectives for the country. In Nijkamp (1996) the following scenarios have been distinguished.

Doomsday scenario

The scenario to be sketched is a troubling one. The world's most populous nation with an extremely tumultuous modern history, still an overwhelmingly rural and poor society in urgent need of industrial modernization would end in cultural revolution.

Despite the efforts for its far-reaching economic reforms, China would remain a socialist, planned economy with a high degree of trade protectionism that clearly exceeds that of any other major country in the region, a country ruled by a totalitarian regime. It is not clear how to incorporate such a large, still predominantly state-directed trading system into a scheme of regional economic co-operation in which the other members' economies are predominantly market oriented. China's options in participating in such an organization would be constrained by historical, ideological and economic factors. The trade partners would impose trade restrictions, such as trade tariffs and quotas, on China. The prospects for further expansion of China's role in the Asia-pacific region would be diminished significantly.

Status quo scenario

The status quo scenario assumes that China would gradually loosen its totalitarian regime. China would need a transition period to implement the outwardly-oriented economic development strategy. However, China's trade regime would still differ from its major trading partners in several respects. Firstly, despite some reduction in the role of central government, foreign trade corporations do not make decisions primarily on the basis of economic criteria; rather, they are subsidised by the government to cover the losses they incur. Secondly, quantitative trade restrictions provide highly variable and difficult to measure protection for domestic producers and would result in a substantial bias against export activities.

As an exception the Schenzhen Special Economic Zone (SEZ), where the market oriented economy has already been developed, will continue to strengthen its economic activities slowly but steadily.

Although China's general policy of economic openness seems unlikely to change abruptly, the results over the long run would depend on the character and pace of domestic economic development.

Heavenly scenario

In the heavenly scenario, the developments of outward-oriented economic policies and the economic reforms would be accelerated, resulting in radical changes in the planned economic system, since the inauguration of Deng Xiaoping's policy at the beginning of the 1980's to modernise China through economic reform and expanding contracts with the West and Japan. The transition to the market economy in China means that projectionist regulations in favour of the national flag and monopolistic practices of the shipping industry towards foreign trade will be gradually abolished. Japan is supposed to liberate trade to the same extent as the regulations in Hong Kong and Singapore did.

The government would embrace capitalism, which makes the future brighter than anyone could ever hope for. Hong Kong would teach mainland China much about free market modes of capitalism and as the mainland "copies" Hong Kong ways, the difference between the two would lessen, resulting in greater stability in the long run. Hong Kong would also extend its links with the mainland more than ever before, utilising the new atmosphere of economic liberalism.

According to the free trade idea, Chinese companies would compete directly with foreign companies, which would encourage better cost management, improved service and greater flexibility. Meanwhile foreign partnerships are being encouraged both to improve the port-management system and raise overall levels of productivity and efficiency. This expansion is attended by increasing trade- and investment relations. Imports of capital and technology would play an important role in sustaining economic reforms.

Clearly, it will be difficult to identify the most plausible image for China. In any case, such a future image can be used as a communication vehicle for testing the robustness of policies and as a learning device for enhancing the flexibility of strategic policies. Next to such broad policy images, there may also be other future developments which impact more directly on spatial processes.

References

Bradbury, I., and R. Kirkby, China's Agenda 21, A Critique, **Applied Geography,** vol. 16, no. 2, 1996, pp. 97-107

Dreborg, K., L. Hedberg, P. Steen, and J. Akerman, Images of Future Transport in Europe, Paper POSSUM project, DG VII, European Commission, Brussels, 1997

Hey, C., P. Nijkamp, S. Rienstra, and D. Rothenberger, Multicriteria Scenario Models for Sustainable Transportation and Regional Development, **Transport Networks in Europe** (M. Beuthe and P. Nijkamp, eds.), Avebury, Aldershot, 1997

Huang, H.J., and Chen, L-Y, Prospects of China's Transportation System, **AIBSEAR '94 (Proceedings of the 10th Academy of International Business)**, Beijing, 1994, pp. 677-682

Kim, T.J., and G. Knaap, The Spatial Dispersion of Economic Activities and Developments Trends in China, Paper European RSA Conference, Rome, August 1997

Klotz, S., and C. Knoth, Regional Impacts of the Open Door Policy in the PR of China, Paper European RSA Conference, Rome, August 1997

Massey, D.W., D. Shaw, and P.J.B. Brown, Economic Imperatives vs. Environmental Quality in the Dragon's Head, Working Paper 56, Dept. of Civic Design, University of Liverpool, Liverpool, 1996

Nijkamp, P., P. Rietveld and H. Voogd, **Multicriteria Analysis for Physical Planning**, Elsevier, Amsterdam, 1993

Nijkamp, P., S. Rientstra and J. Vleugel, Transportation Planning & the Future, John Wiley, Chichester, 1997

Nijkamp, P., Maritime Sector Perspectives of the Pacific Rim, **Studies in Regional Science**, vol.12, 1995, pp. 312-323

Nijkamp, P., and N. Vermond, Scenarios on Opportunities and Impediments in the Asian Pacific Rim, **Studies in Regional Science**, vol. 13, 1996, pp. 1-46

3. Nanjing City and Greater Nanjing

Frank Englmann, University of Stuttgart

3.1 Geographical Location, Area, Population, and Landuse

The city of Nanjing (the old capital of six dynasties of ancient southern China) is located about 300 km to the north-west of Shanghai at the Yangtse-River (31° north latitude, 118° east meridian). Nanjing City is an important traffic node, as it has the biggest river port in China that links Nanjing with the sea. Furthermore, in Nanjing there is located one of the few bridges that crosses the Yangtse River. Hence, it connects Xuzhou (and farther away Beijing) in the north and Zhengzhou in the north-west, Hefei (the capital of Anhui Province) in the west, and Wuxi, Suzhou and Shanghai in the east. Nanjing City is the capital of Jiangsu Province. It has about 2.61 million inhabitants and is therefore one of the biggest cities in China.

Greater Nanjing consists of ten districts and five counties (namely Jiangning, Jiangpu, Luhe, Lishui, and Gaochun) with an area of 6,516 km^2 and 5.22 million inhabitants. Nanjing City consists of the ten districts with an area of 975 km^2 and 2.66 million inhabitants. Six of these districts are urban (namely Xuanwu, Baixia, Jianye, Gulou, Xiaguan), four suburban (namely Pukou, Dachang, Qixia, and Yuhua). The Urban Area consists of the six urban districts with an area of 187 km^2 and 1.8 million inhabitants. (Data as for 1995; see Nanjing Statistical Yearbook 1996 and the maps in appendix A2)

The given data of inhabitants refer to the so called residents. People living and working the whole year in Nanjing City, but without being a registered permanent resident, are called immigrants or floating people and do not belong to the above mentioned inhabitants. 1996 the immigrants amounted to 25 % of the working population.

The Urban Area is confined to that part of the city which lies to the south-east of the Yangtse-River. To the north of the Yangtse-River there are predominantly industrial estates. The Urban Area (or Main City) can be subdivided into 5 parts:
the Northern City (to the north of the railway) with heavy industry, railway stations, and lodgings,

the Eastern City with sights (park, lake, ...), administration, and lodgings,
the Southern City with lodgings,
the Western City, predominantly lodgings, where the number of lodgings is planned to increase strongly in the future,
the Centre (about 60 km^2 and therefore larger than the CBD (Central Business District)) with finance buildings, stores, service stations and lodgings.

3.2 Employment: Its Structure and Location

In the Nanjing Statistical Yearbook the working population is grouped according to various categories:

1. Staff and workers of state-owned entreprises (SOEs), collectives, corporations, joint ventures, and others
2. Employees in urban private entreprises and urban individual labourers
3. Rural labourers (including workers in town-owned firms).

The evolution over time of these categories is shown in Table 3.1:

Table 3.1: Working population in Greater Nanjing according to type of employment

	1987	1990	1995
Staff and workers of state-owned entreprises (SOEs), collectives, corporations, joint ventures, and others	1 418,800 (52.55%)	1,440,900 (51.96%)	1,492,500 (49.96%)
Employees in urban private entreprises and urban individual labourers	23 000 (.85%)	40,600 (2.55%)	134,700 (5.37%)
Rural labourers (including workers in town-owned firms)	1 258,000 (46.60%)	1 261,400 (45.49%)	1,334,500 (44.67%)
Total	2,699,800	2,772,900	2,987,500

Source: Nanjing Statistical Yearbook 1996, p.48

From Table 3.1 one can see that the introduction of market elements into the Chinese economy leads to an absolute and relative increase of the category 'Employees in urban private entreprises and urban individual labourers' at the expense of the two other categories. The overall employment increased by about 10.7% from 1987 to 1995, which means by less than two percent annually. Furthermore, table 1 shows that the total number of employees in category 1 increased by about 5% from 1987 to 1995. This modest increase of less than one percent per year was accompanied by a slow change of employment structure.

Table 3.2: The Evolution of the Structure of Staff and Workers of SOEs, collectives, corporations, joint ventures, and others in Greater Nanjing

	1987	1995
Agriculture	29,100	21,427
Industry	700,600	730,972
Exploration	8,800	11,599
Construction	123,700	111,438
Transportation & Communication	96,900	111,078
Commerce	152,600	166,384
Utilities & Services	63,300	85,725
Hospitals & Sports	28,000	31,288
Education and Culture	111,800	116,891
Science	34,400	39,261
Finance & Insurance	9,500	20,038
Administration	40,400	51,817
Total	1,418,800	1,492,490

Source: Nanjing Statistical Yearbook 1988 and 1996

The number of staff and workers decreased in agriculture by almost 25% and in construction by about 10%. Whereas the growth rate was below average in industry, it was above average in exploration, transport & communication, business, utilities & services, hospitals & sports, science, finance & insurance. Thus, the structural change that took place in Greater Nanjing from 1987 to 1995 can be described as a tertiarization process. This can also be seen from Table 3.3 that shows the distribution of all the working people among the three main sectors in Greater Nanjing.

Table 3.3: Employment in main sectors in Greater Nanjing

	1987	1990	1995
Primary sector	798,800	835,700	791,100
Secondary sector	1,219,300	1,220,900	1,276,100
Tertiary sector	681,700	716,300	920,300

Source: Nanjing Statistical Yearbook 1996, p.48

Most of the staff and workers of SOEs, collectives, corporations, joint ventures, and others are concentrated in Nanjing City, as is shown by Table 3.4:

Table 3.4: The spatial distribution of workplaces in SOEs, collectives, corporations, joint ventures, and others in Greater Nanjing

	1995	1995 (adjusted by multiplying with 7.541524)
Greater Nanjing	1,492,490	
Nanjing City	1,281,253	
Xuanwu	18,021	135,905
Baixia	18,205	137,293
Qinhuai	22,213	167,519
Jianye	17,487	131,878
Gulou	22,752	171,584
Xiaguan	14,112	106,425
Pukou	15,235	114,895
Dachang	13,883	104,698
Qixia	16,727	126,147
Yuhua	11,258	84,902
Rural Nanjing	211,237	
Jiangning	56,699	
Jiangpu	23,318	
Luhe	47,852	
Lishui	43,520	
Gaochun	34,848	

Source: Nanjing Statistical Yearbook 1996, p.47

Unfortunately, the numbers for the ten districts do not add up to the number for Nanjing City (the sum over the ten districts is 169.893, instead of 1.281.253). By assuming that the relative values for the ten districts are correct and that the number for Nanjing City as a whole is also correct, the adjusted numbers can be derived. The low number of workplaces in SOEs, collectives, corporations, joint ventures, and others in Rural Nanjing corresponds with the comparatively low numbers of non-agricultural population there. See the following table.

Table 3.5: The spatial distribution of rural and non-rural population in Greater Nanjing

	Non-rural population	Rural population
Greater Nanjing	2,590,417	2,626,778
Nanjing City	2260,187	397,791
Xuanwu	332,421	16,611
Baixia	283,453	11,068
Qinhuai	210,390	11,595
Jianye	214,058	8,592
Gulou	437,201	6,324
Xiaguan	263,257	1,863
Pukou	103,269	56,598
Dachang	145,689	31,323
Qixia	158,926	161,102
Yuhua	111,523	92,715
Rural Nanjing	330,230	2,228,987
Jiangning	100,127	639,991
Jiangpu	51,158	247,375
Luhe	89,359	594,158
Lishui	45,096	361,476
Gaochun	44,490	385,987

Source: Nanjing Statistical Yearbook 1996, p.23

In Nanjing City the rate of unemployment was below 3% in the years 1990-1995. According to the 9th five-year plan it is supposed to remain below 4% in the years 1996-2000.

3.3 Industrial Development Zones

In Greater Nanjing nine industrial development zones were created in order to concentrate industrial firms there. These nine development zones are located outside of the traffic cells. Hence, they have to be accounted for by the various directions. Furthermore, it must be noted that the influence of an industrial development zone on the traffic volume of any of these directions is the smaller, the farther away this industrial zone is located from the area covered by the traffic cells. Hence, for the traffic situation in the area covered by the traffic cells, those industrial development zones are most important that are located nearby the feeding points of the various directions. For our purposes the organizational structure of these industrial

development zones is not important, only their location (see also Map 2 in appendix A2 at the end of the report)
According to these criteria, the following are to be considered in more detail:

1. The Headquarters of the Nanjing High and New Technology Industry Development Zone, located in Pukou, i.e. in the vicinity of the feeding point of direction R1 (see Maps 2 and 3, appendix A2).
2. The Nanjing Economic-Technical Development Zone including the Nanjing Xingang High and New Technology Industry Park, located in Qixia, in the vicinity of the feeding point of direction R2 (see Maps 2 and 3, appendix A2)
3. The Nanjing Jiangning Economic and Technological Development Zone including the Nanjing Jiangning High and New Technology Industry Park and the Nanjing Non-State-Run Science and Technology District, located in Jiangning county, i.e. outside of Nanjing City. Still, their location is not too far away from the feeding point of direction R5 (see Maps 2 and 3, appendix A2).

Ad 1:
The planned area is 6.5 km^2. Here mostly R&D departments will be located. This includes pre-production and only to a minor extent production. Hence, the freight traffic is only of minor importance, whereas the density of work places is relatively high.

Ad 2:
The planned area is 9.73 km^2. The main industrial products are computers, modern communication equipments, pharmaceuticals, new materials, and fine chemicals. Compared to the areas in Pukou the density of work places is lower, whereas the share of freight traffic in total traffic generated by the industrial zone is higher. But with respect to freight traffic it has to be kept in mind that the industrial estate is in close vicinity of both a port facility and a railway station.

Ad 3:
The planned area is 24.98 km^2, out of which 70% are designated for the production of goods and services, the rest for residential and recreational purposes. It is planned to turn the development zone into an urban district. Hence, the chances are good that a big share of the employees will reside within the development zone or its vicinities. Furthermore, the ringroad around the urban area of Nanjing that is under construction also has to be

taken into account, as this will be the major link of the Nanjing Jiangning Economic and Technological Development Zone and the port.

The amount of passenger traffic generated by the location of work places in the vicinity of the feeding points of the various directions heavily depends on where the employees will be living. If they are going to live in the area covered by the traffic cells, this will heavily add to the number of work trips in the respective directions. Instead, if they are going to live close to their work places this will only generate some leisure or shopping trips. Furthermore, the amount of freight traffic generated depends on the transport intensity of the production process and on the composition of the firms located in a development zone. If the various production stages are located within a development zone this reduces the amount of freight traffic generated outside the development zone. As can be seen from the brochures of the various development zones by which they try to attract investment, this is one of the guiding principles for the sectoral composition of development zones in Greater Nanjing.

3.4 Income, Expenditures, and Availability of Durable Consumer Goods

Table 3.6 shows that the share of expenditures for transport and communication is still very low in Nanjing, compared to Western European standards. This is due to the fact that per capita income is still very low, too. Hence, also the car ownership rate is very low. The latter is further reduced by the high prices of cars, even if they are produced in China. The cheapest middle class car produced in China costs about 140,000 yuan, the yearly car tax is 20,000 yuan. The latter is the expression of a policy in China to keep the business car ownership rate low. This increases the prices for private cars as well, even if the central government decides to pursue a policy to increase private car ownership. Hence, there exist very important economic reasons, why the most important passenger traffic modes are walking, biking, and public transport: these are rather inexpensive (compare with section 4.1). Furthermore, the city government of Nanjing specifically restrains the number of permits for private and government cars, less so for business cars, in order to restrict the volume of motorized individual traffic.

Table 3.6: Per capita income, savings, and expenditures per capita

	1994	1995
Monthly net income per capita	332.47 yuan	416.34 yuan
Monthly expenditures per capita	293.08 yuan	376.99 yuan
Savings rate	11.8%	9.5%
Share of expenditures for food and beverages in total expenditures	52.85%	53.39%
Share of expenditures for clothes and shoes in total expenditures	13.36%	12.88%
Share of expenditures for transport and communication in total expenditures	5.84%	4.06%
Share of expenditures for travel in total expenditures	.96%	.85%
Share of expenditures for rent in total expenditures	1.16%	1.07%
Share of expenditures for water charges in total expenditures	.40%	.41%
Share of expenditures for electricity in total expenditures	1.68%	1.68%

Source: Nanjing Statistical Yearbook 1996, p. 308f.

This also becomes clear from the following table that shows the availability of durable consumer goods in families of Nanjing City. (The average family size is 3.10 persons.)

Jiangsu Province is the province with the second highest gross domestic product (GDP) per capita in China. In the year 1997 an increase of nominal GDP by 13 % is expected for Greater Nanjing. An increase of 22.6 % per year is planned until the year 2000. The inflation rate is supposed to be less than 10 %. Hence, the yearly growth rate of real GDP is planned to be higher than 12 percent. In the years 1996 - 2000 a yearly average increase of wages in SOEs of 14.8 % is expected. This would imply that nominal wages will almost double. But, still, the wages would be so low on average that the (private) car ownership rate would only slowly increase in absolute terms, even if the percentage rate of growth may be high due to the low initial level. Furthermore, the latter development presupposes that the city government of Nanjing would alter its restrictive policy concerning the permits for private cars.

Table 3.7: Availability of durable consumer goods per 100 families in Nanjing City

	1995
Bicycle	226.0
Motor bike	1.7
Launderer	95.3
Refrigerator	92.3
Colour TV	95.3

Source: Nanjing Statistical Yearbook 1996, p. 311

3.5 The Housing 'Market'

Table 3.6 shows that the expenditure share for rent in 1995 is surprisingly low. This is at least partly due to the organization of the housing 'market'. In Nanjing City two market segments have to be distinguished:

1. The government segment which consists of houses directly or indirectly owned by the government. Mainly, these houses or appartments are rented to present or former employees. The respective rents are low, irrespective of the location of the lodgings. Here the low rents can often be considered as a hidden part of income.
2. The market segment which consists of houses owned by housing corporations and private persons: The respective rents are determined by demand and supply, which means that here location matters. The rents are higher in the centre than in the periphery.

The share of the first category is by far bigger than that of the second one, as is also indicated by the low expenditure share for rent. Another reason for this low expenditure share is the small floor space per capita in Nanjing City. It was only 8.36 m^2 in 1996, for the year 2000 the floor space per capita is planned to be 10 m^2. (In the rural counties floor space per head is about 25m^2). This shows that the construction of lodgings is one of the important sectors of the 9th five-year plan (1996-2000). According to this plan, the emphasis will be given to increasing the market segment.

4. Traffic Development in China (especially within the Region of Nanjing)

The actual transport situation of the P.R.China has to be seen against the following background:

- the historical strong position of railway politics (investment- and price policy),
- the very long distances for transportation with domination of raw material and
- agricultural products,
- the shortage of high-quality streets,
- the shortage of all-year passable and efficient connecting streets between the different provin-ces,
- the absence of a sufficient supply and service system for motor vehicles.

With respect to the transport of goods on the roads essentially goods such as coal, building materials, fertiliser and cereals are transported in the local zone. The typical size of a consignment is the payload of one single lorry. The restricted volume of the railway traffic causes the transportation of goods along the street traffic, even at long distances.

A proportion of bus routes lead to neighbouring districts, some also lead to other provinces. In all districts where no railway and no navigation exist, the bus traffic is the only traffic mode available.

4.1 The Traffic Development until 1996

As most Chinese cities since the opening of China, 1978, Nanjing is also exhibiting a rapid development in industry, the tearing down and building up of lodgings and factories, the standard of living and especially the evolution of the traffic and transportation sector. This rapid development can be illustrated by the traffic within Nanjing:
During the last 40 years, the total number of vehicles in the Urban Area increased in an almost li-near way, whereas the number of bicycles has almost reached satiation: 1.6 million (for 1.95 million inhabitants).

At present the total length of streets in the Urban Area is 1.033 km: 113 km primary streets, 63 km secondary streets. The public traffic system transports 444 million persons every year. The average velocity of all cars of Nanjing amounts about 16 km/h.

On the one hand the traffic volume of Nanjing increases more than 15% per year, on the other hand the length of streets increases only 5% per year (e.g. during the last 10 years about 40 streets have been broadened). Therefore, the traffic jams increase in number and duration.

While in the years 1950 - 1980 the number of trucks saw the greatest increase, today all vehicles of the individual traffic have reached the top of stock development. In 1987, the number of motor-cars per 1000 persons was 0.3 (comparison with industrial nations: 200 - 550 motor-cars per 1000 persons).
In conclusion, it can be said that in an overall view over all Chinese provinces, Jiangsu province is lying near the average with respect to the motorization in these years.

Before looking at traffic development until 1996 with respect to several aspects, one should briefly analyse the financing of road construction:

Usually, several administration units take part in financing streets. National streets are not only financed by the central government but also by the provinces. The latter participate with approximately 50% of the costs. The lower administration units are also sufficiently short of financial funds so that they are not able to finance the streets on their own. Therefore, the authorities of economic and industrial importance support them in this sector.

The municipal streets are built by the population itself at very low cost, with the support of the government.

The financial funds of the investments have different sources. Firstly, the motor-vehicle tax and the street charges are revenues of the Chinese Ministry of Transport. Secondly, general budgetary means are used for road construction projects. 1987, for example, the street charges were 10 billion Yuan. They were divided into 8 billion Yuan for street maintenance and 2 billion Yuan for constructing new roads.

As a further possibility on a small scale the road construction is financed by credits. Here, the repayment is secured by charging a toll for using streets or bridges. This may be an important fact for Nanjing City because of its one single bridge over the Yangtse-River in the north-west. For some projects the financing can be provided via foreign credits by the national government.

In details, the traffic development of the recent 10 years is as follows:

1. Numbers of vehicles:

The most remarkable aspect of the numbers of vehicles in China is the very large preponderance of bikes. In Nanjing City the bikes amount to more than 95% of all vehicles. On the other hand the number of bikes has already nearly reached a state of satiation. Today, more than 80% of the total population own a bicycle. The increase from 1995 to 1996 amounted only to 2%, whereas the yearly average increase of the last ten years was about 6%.

Table 4.1: Numbers of vehicles

Category / Year	Bikes	Buses	Private cars	Firms cars	Trucks
1987	1201299	1117	85	9335	21745
1988	1332761	1127	273	10546	23792
1989	1466991	1128	751	11317	24569
1990	1583099	1131	1388	11884	25469
1991	1171377	1170	1596	13255	26813
1992	1344956	1316	1915	13906	28837
1993	1523962	2412	2490	21529	33435
1994	1659769	2469	3112	26187	35236
1995	1846468	2321	3758	29998	38181
1996	1886667	2328	4698	35108	41961

Source: Nanjing Yearbook (1987-1996) and unpublished documents of the Traffic Police Detachment of Nanjing, China

There exist two main reasons for this preponderance of bikes:
Firstly, of course the P.R.China has been - and still is - an emergent nation. The economic situation did not allow most people to afford a private car.

Secondly, one of the few principles in letting lodgings in Nanjing is that there is a short distance between the lodging and the factory where the tenant works. Therefore, at least for the trips between home and working place, private cars were not badly needed.

(The decrease of more than 400.000 bikes in 1990 may be the result of a correction in the Chinese statistics which is beyond our access.)

In 1987, the ratio between firms cars and private cars was 110:1. With a considerable improvement of the standard of living especially the number of private cars increased from 85 cars (1987) to 4689 cars (1996) what means an increase of more than the 50-fold in ten years. Furthermore, the growth is accelerating. For the time being, there is no end to this development in sight. A possible long term ratio of cars per person could be 1:1 all over the world and therefore even in China. Forecasts should include this possibility in their scheme of scenarios.

Notice, that at present still only 0.2% of the population have a car on their own. On the other hand the desire for possessing a private car is very strong in China, too, so that people will buy a car as soon as they are able to afford it.

The bus situation is much less dramatic. In the years 1987 - 1991 there was no change of the stock of buses. The great difference in the number of buses between 1992 and 1993 may be caused by an extension or modification of the Chinese statistics as compared to the years 1979 and 1984. From 1994 to 1995 the bus stock even decreased. In the same year the number of bikes increased at the highest rate within the last decade. At present, there are twice as much buses than private cars. But as mentioned above this ratio will not hold for long.

The number of firm cars which make up with over 99% at present the large majority of all existing cars in Nanjing has also increased with a yearly average growth rate of 17%.

Finally, the trucks should be mentioned with a yearly average growth rate of 8%. It can be foreseen that in a few years the number of firms cars will exceed the number of trucks.

In total, the number of all vehicles in Nanjing City in 1996 is about the 1.6-times the number in 1987, whereas the number of all vehicles *except for bikes* in Nanjing City in 1996 is more than the 2.6-times of the number in 1987.

2. Travelling Costs:

a) Travelling Costs for the Urban Bus:

(The data are obtained from unpublished documents of the Commodity Office of Nanjing, 1996)

In Nanjing City, the tariff structure works as follows:
The bus traveller has to pay a basic amount to start the trip and an extra amount for each bus stop. In addition a monthly ticket is available. In 1987 the price for a general ticket was 0.05 Yuan (today about 0.01 DM) to start and also 0.05 Yuan for each stop. Approximately every two years all bus prices are raised so that 1989 the starting price was already 0.1 Yuan, but again 0.05 Yuan for each stop whereas in 1996 the starting price was 0.4 Yuan and 0.2 Yuan for each stop. This means an average growth rate of 75% for the starting price and of 50% for the price for each bus stop. On the other hand a monthly ticket has been getting relatively cheaper in the last ten years with an average growth rate of only 18%.

Arguing for example on the assumption that a trip includes five bus stops this would mean:
In 1987 one had to make 22 trips to let a monthly ticket of 6.5 Yuan be cheaper whereas in 1996 the price of a monthly ticket of 25 Yuan corresponded to only 18 trips.

In general, the costs per person and kilometre increased from 1987 (0.015 Yuan) to 1996 (0.085 Yuan) with a yearly average growth rate of 55% which means an unusually high increase in the prices for urban buses.

b) Travelling Costs for the Train:

(The data are obtained from the Price list of Passenger Tickets (Railway Bureau of China)

The consideration of the costs for the train is a little bit more complicated because one has to distinguish between several categories. Until

1995 there existed three different types of trains: a general train and two types of express trains. Since October 1995 the express trains are also available with air conditioning. Another subdivision is made with respect to comfort. In every train two kinds of seats exist: hard and soft.

In addition, the costs for these categories are dependent on the length of the trip. This subdivision includes four categories:

- 1 - 20 km
- 21 - 30 km
- 31 - 40 km
- 41 - 50 km

Since 1987 only two price corrections were made. The first in October 1989, the second in October 1995. Generally speaking, all prices range between 1 Yuan (for a short trip on a hard seat in a general train) and 8 Yuan (for a long trip on a soft seat in a fast express train with air conditioning).

By the way, the difference between the prices for a hard seat and a soft one amounts in all categories and for all lengths under 30km to 1 Yuan, and for longer distances to 2 Yuan. The following observations are for hard seats:
Beginning with the general train the price span between short and long distances is exactly 2 Yuan. Here, the prices increased in October 1995 by 0.5 - 1 Yuan (33% - 100%) for the longer distances whereas the price for the short distance stayed the same for the last 10 years.

The price spans of the express trains of the types A and B are currently 2 Yuan (type B: 2 Yuan for the short distance, 4 Yuan for the long one; type A: 3 Yuan for the short distance, 5 Yuan for the long one). The increase ratio of the prices in October 1995 is similar to the case of the general train.

The most expensive way to travel by train is to choose express trains with air conditioning with prices which lie exactly 1 Yuan over the prices in comparison to trains without air conditioning. This holds for all lengths of the chosen trip.

In this context, it should be mentioned that for the year 2020 a high-velocity-line with a veloci-ty of 250 km/h (at present 170 km/h) is planned for the outskirts of Nanjing.

c) Travelling Costs for Ships:

(The data are obtained from the Port Office of Nanjing, China)

Since the Yangtse-River flows directly through the Urban Area of Nanjing, ships plays a role in the traffic development of Nanjing even though this role should not be overrated.

The existing data refer to the prices for the travelling costs from one dock to another. These prices are distance-dependent. Only on the route from Shagan to Pukou does there exist a quick boat (since 1995).

The adjustment times for prices were May 1989, June 1992 and the last one in June 1995. In distance-dependent order each possible pair of two docks within Greater Nanjing is listed below

- Shagan - Pukou
- Shagan - Nanchangmen
- Yanziji - Tongjangji
- Shagan - Yanziji
- Shagan - Zhenjang
- Shagan - Caishiji

The first three connections all cost between 1 and 1.7 Yuan with average growth rates of 59%, 64% and again 64% in the three above mentioned adjustment times. The current price for the quick boat (between Shagan and Pukou) is 1.5 Yuan.

The last three connections are with prices between 8 and 12 Yuan more expensive than the first ones. They had average growth rates of 54%, 52% and finally 36%. All other routes had in 1996 a standard price of 0.3 Yuan with an increase of nearly always 100% per price adjustment.

d) Travelling Costs for the Taxi:

(The data are obtained from unpublished documents of the Commodity Office of Nanjing, 1996)

The tariff structure concerning the travelling costs for taxis in Nanjing again is a little bit more complicated because of the several categories and the formalism of the distance-dependent pay structure. The idea is the same as in many other countries:
The traveller has to pay a base price which is the price for the trip within a specified distance, the so-called base length. For each additional kilometre the traveller has to pay a certain price per km. In the course of the last 10 years all three parameters

- base length
- costs of the base price
- price per each additional kilometre

changed four times (in April 1986, in December 1988, in October 1991 and finally in 1994)

In addition to that there exist three different car sizes

- small
- medium
- large

Remarkable is the very large increase of the taxi price in December 1988:
The costs of the base price for a small car grew from 2.20 Yuan to the 4-fold (8.80 Yuan). For medium-sized taxis the price increase ran up to almost 30% while the corresponding base length increased from 8 to 10 kilometres.

It is true that the prices for large taxis increased in the same year by more than 65% but on the other hand the base length also grew by exactly 50% from 10 to 15 kilometres.

All prices had a second large increase in October 1991 while the corresponding base lengths essentially remained constant. The prices for small taxis increased by about 36% with respect to all parameters; for medium-sized taxis by 57% with respect to the base price, 20% with respect to the price per kilometre; for large taxis by 89% with respect to the base price and finally 26% with respect to the price per each additional kilometre.

In 1994 it is remarkable that the prices were decreased for small taxis while all other prices stayed constant. The current price situation for taxis is the following:
For small cars the base price is 8 Yuan with 3 km for free and 2.67 Yuan for each additional kilometre. For medium-sized cars the base price is 21 Yuan with 10 km for free and 2.10 Yuan for each additional kilometre. For large cars the base price is 60 Yuan with 15 km for free and 4.00 Yuan for each additional kilometre.

4.2 Trends and Scenarios in the Transportation Sector

The central organizational obstacles to an efficient division of labour in the transportation sector in the P.R.China have their origin in the national planning and disposition sector as well as in the traffic authorities.

Up to now the planning for the transportation infrastructure of railway, road transport and inland navigation are done to a great extent separately. But an efficient use of small resources requires integrated planning. The separation of the administrative and planning competences for the railway and all other transport modes is a handicap for such integrated transportation planning. Therefore, a combination of the competences in one ministry with several integrated planning and co-ordinating departments would be helpful.

In the actual economic development phase, China needs a national planning program with respect to the planning of the traffic network as well as of the whole transportation sector.

The following measures are planned until 2020 in Nanjing:
In order to unburden individual traffic, three underground railways are planned and already accepted: for the years 2000, 2010 and 2020. But presumably, Nanjing is short of the necessary money. Since there is only one single bridge over the Yangtse-River in the north-west of Nanjing and since the erection of other bridges is very expensive, the planning commission proposed to shift the main city traffic to the south and the south-east

of Nanjing. This strategy is also of interest because the new airport (finished in 1998) is about 20 kilometres south-east of the Urban Area of Nanjing. This new airport is used for international flights, whereas the old one will be at the dispose of domestic flights.

Furthermore, for the year 2020 it is planned to connect the environs of Nanjing with a high-velocity-line of a velocity of 250 km/h (at present 170 km/h). For the years 2020 - 2050 another high-velocity-line with a velocity of about 200 km/h is planned which shall connect the cities Beijing in the north, Shanghai in the east and Hong-Kong in the south. This would underline Nanjings role as the connecting link in eastern China.

Part II

Description of the Model and of the Adaptation to the Available Data

5. The Integrated Transport and Evolution Model

The impact of transport infrastructure on regional development has been difficult to verify empirically in Europe (Wegener in WP 7, EUROSIL, 1998). There seems to be a clear positive correlation between transport infrastructure endowment or the location in interregional networks and the *levels* of economic indicators such as GDP per capita (e.g. Biehl, 1986; 1991; Keeble et al., 1982, 1988). However, this correlation may merely reflect historical agglomeration processes rather than causal relationships effective today (cf. Bröcker and Peschel, 1988). Attempts to explain *changes* in economic indicators, i.e. economic growth and decline, by transport investment have been much less successful. The reason for this failure may be that in countries with an already highly developed transport infrastructure further transport network improvements bring only marginal benefits. The conclusion is that transport improvements have strong impacts on regional development only where they result in removing a *bottleneck* (Blum, 1982; Biehl, 1986; 1991). In China the situation is quite different. There is a growing need for an expansion of transport infrastructure on the urban level, the regional and national levels.

While there is uncertainty about the magnitude of the impact of transport infrastructure on regional development, there is even less agreement on its direction. It is debated whether transport infrastructure contributes to regional polarisation or decentralisation. Some analysts argue that regional development policies based on the creation of infrastructure in lagging regions have not succeeded in reducing regional disparities in Europe (Vickerman, 1991a), whereas others point out that it has yet to be ascertained that the reduction of barriers between regions has disadvantaged peripheral regions (Bröcker and Peschel, 1988). From a theoretical point of view, both effects can occur. A new motorway or high-speed rail connection between a peripheral and a central region, for instance, makes it easier for producers in the peripheral region to market their products in the large cities, however, it may also expose the region to the competition of more advanced products from the centre and so endanger formerly secure regional monopolies (Vickerman, 1991b; Bundesminister für Verkehr, 1996).

Aschauer's (1989) empirical examination confirmed the significant positive relation between governmental infrastructure and factor productivity. The connection between infrastructure and economic growth was even confirmed at a regional level. However there are lags; not all growth effects of infrastructure can be measured locally. The narrower the focus of research, the more the effect decreases. The output elasticity of the public capital stock for the states of the USA was found to be 0.15, only half of the national level (Munnell, 1993). Munnell disaggregated his research by regions, industries and urban areas. Regions with a better public capital stock had higher individual output, too. The main net effects could be attributed to road investments and water supply. Other regional approaches used a production function to assess the influence of infrastructure on development (e.g. Schalk, 1976; Blum, 1982; Rietveld, 1989; Garcia-Mila and McGuire; 1992, Rousseau et al. 1992).

Based on the public-capital hypothesis, the connection between infrastructure and development is measured directly. As an alternative it was proposed to assume a supplementary relation between infrastructure and production. This allows the influence of the infrastructure to be modelled in a more differentiated way (Berndt and Hansson, 1991). Seitz et al. calculated the reduction of production costs by infrastructure (Conrad and Seitz, 1992; 1994; Seitz and Licht, 1993; Seitz, 1994; 1995). They found that the increase in productivity with public capital was higher than without. In conclusion, public capital such as transport infrastructure can partially explain the development of economic productivity. However, in many studies this result could not be confirmed (Erber, 1995, Hofmann, 1996; Kitterer and Schlag, 1995; Kitterer, 1998).

The important role of transport infrastructure for regional development is one of the fundamental principles of regional economics. In its most simplified form it implies that regions with better access to the locations of input materials and markets will, *ceteris paribus*, be more productive, more competitive and hence more successful than more remote and isolated regions (see Linneker, 1997).

According to SACTRA (1998) a list of important regional effects of transport investment effects has to include the following aspects: Transport investment may broaden the access of employers to qualified labour, expand market areas, attract inward investment, improve the image of a region, unlock suitable development sites and induce further economic activity and further employment. However, there may be also negative impacts: The net effect on employment and regional activities depends on the balance between export promotion and import substitution for local production. Transport improvement may have displacement effects in other regions. Marginal changes in the quality of an already efficient infrastructure system are less likely to have significant effects. Transport investments may reduce the demand for transport resources (e.g. drivers and vehicles) by improving the productivity of the transport sector. And finally, labour market characteristics have to be considered. Transport investment can generate positive employment effects at three levels:

(a) *Direct employment effects of investments in transport infrastructure.* The importance of employment effects during the building period is small. After the end of the investment phase most of the employment effects disappear. Some of the employment generated is not necessarily established in the catchment area but in other regions (e.g. employment through construction or planning). The direct employment effects of transport investment have been measured by multiplier analysis (Schmidt, 1976; Baum, 1982).

(b) *Employment effects in the automobile industry, transport-related industries and other industries.* Using input-output-analysis the effects of infrastructure investments on automobile production and transport related industries can be determined. Secondary effects will be initiated by primary employment effects because of changes in consumer behaviour. However, not only losses of income due to production losses in the car-related industries and their sub-suppliers as well as the other affected consumer goods industries are to be expected but also productivity changes in all industries due to reduced division of labour as a result of mobility changes. While employment-related input-output analysis is often used in sectoral economic research, its application in transport and labour market economics is yet rare (Baum, 1982; DIW, 1990, DIW, 1992). An important extension of input-output analysis to a general equilibrium model is the Australian ORANI simulation model (BTCE, 1996). It reduces the shortcomings of input-output-analysis by including e.g. government deficits, balance of trade, availability of resources and changes in prices. An ORANI simulation for Australian highways shows significant employment effects of transport infrastructure investments.

(c) *Employment effects due to location advantages and increased growth.* The employment effects due to increased growth prevail after the investment phase. In the economic literature these effects are discussed intensively and with opposing opinions (Lutter, 1981; Aberle, 1981; European Commission, 1997b; SACTRA, 1998; Zachcial, 1998).

The discussion about the importance of infrastructure capital for economic growth was revived at the end of the 1980s by the so-called public-capital hypothesis. Pioneered by Aschauer (1989), the hypothesis states that increases in public capital, i.e. public investments, will have either positive or negative (crowding-out) influence on private investment and productivity. One part of economic capital is directly linked to the transport sector. Public infrastructure capital is a part of the whole capital stock, so increases in public infrastructure will generate private investment.

The discussion about the public-capital hypothesis produced a great number of publications that often confirmed the importance of transport-related investment on economic development (Hulten and Schwab, 1991; Ford and Poret, 1991; Neusser, 1992; Rebelo and Easterly, 1993; Baffes and Shah, 1993; Munnell, 1990; 1993; Canning and Fay, 1993; Uchimura and Gao, 1993; World Bank, 1994). However, the hypothesis has also been criticised because of incompleteness and insufficient modelling (Aaron, 1990; Munnel, 1993; Gramlich, 1994).

Travel demand can be seen as a result of economic and individual activities of firms and/or individuals, and therefore of individual decision processes of the actors. Therefore, the traffic volume is influenced by the total number of activities, this means by the places and times of activities, the modes of transport and the route selection, the co-ordination of activities with other actors (e.g. formation of car pools) as well as by feedback effects.

The decision processes of the individuals or households to travel (micro level) are reflected in the traffic flows between the traffic cells or the traffic volume (macro level). The complex decision behaviour of the individuals is represented and analysed by a dynamic non-linear decision model. The model provides a transition from the micro level (decisions of the individuals) via a statistical procedure to the macro level (resultant traffic flows).

Proceeding in this way it is possible to find out by regression analysis the relevant key factors of the travel decisions from the empirically amenable traffic flows (number of the trips between the traffic cells) and the respective population numbers.

Accessibility is the main 'product' of a transport system. It determines the locational advantage of a region relative to all regions (including itself). Indicators of accessibility measure the benefits households and firms in a region enjoy from the existence and use of the transport infrastructure relevant for their region.

The travel decisions can be influenced by different factors like travel times or accessibility of the traffic cells, the housing or labour market. The key factors and their influence on the travel behaviour can thus be systematically determined. Changes in the key factors lead to changes in the traffic volume and distribution, because of the reactions of the decision makers to the changed general conditions. In this way, the basic requirements for a sensitive forecasting model are fulfilled. The model is represented in detail in the appendix.

5.1 Achievements of the Model

The decision based ITEM transport model simultaneously provides traffic generation, traffic redistribution and modal split. Individual decision processes of agents are the basis of all transport phenomena. Therefore, the trip frequency is the result of a great number of "individual" decisions and is related to the rational and irrational motives of the agents. On the basis of the master equation approach (Weidlich and Haag, 1983) the probability of finding a certain trip distribution taking into account the links and feedback's between the decision processes involved, can be calculated. The most probable trip distribution is then related to empirical trip matrices, in order to estimate the parameters of the ITEM-transport model. Intermodal as well as multimodal effects may be incorporated within a wide range of mathematical functions.

The ITEM transport model offers a series of advantages as compared to traditional analysis and forecasting methods:

- The description is dynamic, i.e. the evolution with time is represented.
- The model includes non-linear effects, i.e. interactions within the population are considered (e.g. imitation or anti-imitation effects). As a consequence, the interactions between the traffic cells are also represented.
- The immense amount of information contained in the empirical traffic flows is reduced to only a few well interpreted parameters like attractivities, resistance parameters and mobility indicators. These parameters are described in the following section.
- In a first step of the estimation procedure the attractivities are found from the empirical data base. In a second step the key factors determining these attractivities are calculated by regression analysis. In tra-

ditional models, the attractivity fa:ctors are often only given as external input.

- The ITEM transport model as a dynamic model can describe the spatial redistribution of individuals over the day, i.e. their daily route chains, are represented in aggregated form. This is a decisive difference to traditional traffic models which are not able to describe the variation over the day. Since the number of travel decisions highly depends on the actual distribution of the actors in the respective cells, the consideration of the redistribution of the population is not only plausible, but improves simultaneously the model quality.
- The traffic generation and the traffic distribution are considered in one single step.

The ITEM transport model represents therefore a suitable extension of the usual methods for the analysis of traffic behaviour and traffic planning.

5.2 Model Description

Individual transition probabilities between the traffic cells are the essential components of the ITEM transport model. These transition probabilities represent the probability that a representative individual of a traffic cell travels into another cell within a specific period of time for a specific trip purpose using a specific transport mode. The individual can be assigned to one of a variety of different groups which can be distinguished e.g. by age or gender or profession. Each of these groups exhibits a rather uniform probabilistic decision behaviour.

The modelling of the transition probabilities with the aim of representing the traffic behaviour of the individuals is the central task. Time dependent attractivity differences between the traffic cells and time-dependent resistances (distance, travel time) between the traffic cells are considered to be the essential influencing factors of the traffic flows. Moreover uncertainty within the decision process e.g. because of insufficient information is included in the model. For instance a household is normally much better informed about the purchase possibilities in the local environment than about purchase possibilities in traffic cells sited far away. Therefore these purchase possibilities far away are of only subordinate influence in the household decisions.

Applications of this transport model for the analysis and simulation of traffic behaviour in the Fe-deral Republic of Germany show, that three factor sets (indicators) essentially come into conside-ration as determinants of those transition probabilities (The Federal Minister for traffic, 1996):

- *„Attractivities"* of the traffic cells for a travel decision which depend on a bundle of different key factors. *Attractivity differences* can be seen as *„driving forces"* for the travel decisions. I.e. the larger the attractivity difference between two traffic cells, the more probable becomes a trip into the more attractive traffic cell. Simultaneously a trip from the more attractive traffic cell into the less attractive one becomes less probable.
- *Resistance parameters* for the trip between the traffic cells which describe the influence of travel times, travel costs or also of the comfort of a specific transport mode.
- *Mobility indicators* of the population. This parameter describes the average number of the trips relative to the entire number of inhabitants, in the case of a given attractivity difference between the traffic cells. Therefore a duplication of the mobility indicator leads directly to a duplication of all traffic flows between all traffic cells. The mobility indicator is directly correlated with the trip frequency and the mobility behaviour. The indicator is time-dependent and depends on trip purpose and transport mode.

In this way, the information contained in the traffic flows can be reduced to a few relevant and well interpretable factors.

The attractivity of a traffic cell depends on the one hand on the population distribution, and on the other hand on a set of socio-economic variables such as e.g. the distribution of the work places or of the housings.

Following the definition and interpretation of Wegener (1994, 1998a) accessibility indicators can be defined to reflect both *intra*regional transport infrastructure and *inter*regional infrastructure which affect the specific region.

Simple accessibility indicators consider only intraregional transport infrastructure expressed by such measures as total length of motorways, number of railway stations (e.g. Biehl, 1986; 1991) or travel time to the nearest nodes of interregional networks (e.g. Lutter et al., 1993). While this kind of indicator may contain valuable information about the region itself, they fail to recognise the network character of transport infrastructure linking parts of the region with each other and the region with other regions.

More complex accessibility indicators take account of the connectivity of transport networks by distinguishing between the network itself, i.e. its nodes and links, and the 'activities' (such as work, shop or leisure) or 'opportunities' (such as markets or jobs) that can be reached by it (cf. Bökemann, 1982). In general terms, accessibility then is a construct of two functions, one representing the activities or opportunities to be reached and one representing the effort, time, distance or cost needed to reach them:

$$A_i = \sum_j g(W_j)\, f(c_{ij})$$

where A_i is the accessibility of region i, W_j is the activity W to be reached in region j, and c_{ij} is the generalised cost of reaching region j from region i. The functions $g(W_j)$ and $f(c_{ij})$ are called activity functions and impedance functions, respectively. They are associated multiplicatively, i.e. are weighted to each other. That is, both are necessary elements of accessibility. A_i is the accumulated total of the activities reachable at j weighted by the ease of getting from i to j.

It is easily seen that this is a general form of potential, a concept dating back to Newton's law of gravitation and introduced into regional science by Stewart (1947). According to the law of gravitation the attraction of a distant body is equal to its mass weighted by a decreasing function of its distance. Here the attractors are the activities or opportunities in regions j (including region i itself), and the distance term is the impedance c_{ij}.

Different types of accessibility indicators can be constructed by specifying different forms of functions $g(W_j)$ and $f(c_{ij})$.

The interpretation here is that the greater the number of attractive destinations in regions j is and the more accessible regions j are from region i, the greater is the accessibility of region i. This definition of accessibility is referred to as destination-oriented accessibility. In a similar way an origin-oriented accessibility can be defined: The more people live in regions j and the more easily they can visit region i, the greater is the accessibility of region i. Because of the symmetry of most transport connections, destination-oriented and origin-oriented accessibility tend to be highly correlated.

The concept of daily accessibility is due to Törnqvist who in 1970 developed the notion of 'contact networks' hypothesising that the number of interactions with other cities by visits such as business trips would be a good indicator of the position of a city in the urban hierarchy (Cederlund et al., 1991). In the accessibility study of the BfLR for DG XVI (Lutter et al., 1993) daily accessibility was calculated in terms of the number of people that can be reached in three hours by the fastest mode. Modes considered included road, rail and air with and without planned infrastructure investments (new motorways, high-speed rail lines and more frequent flight connections).

Within ITEM it seems to be plausible to introduce accessibility indicators for the daily home-work trips e.g. number of accessible work places within 30 minutes instead of a 3 hour limit typical for business trips.

The influence or the significance of these variables (attractivity factors) is determined by means of a multivariate regression. The composition of the relevant variable sets obviously depends strongly on the trip purpose. In general the attractivities of the traffic cells can be set in relation to the following different variables of the individual cells (the Federal Minister for traffic 1996):

- number of the inhabitants or neighbours reaching the cell within 30 minutes
- number of employees or of attainable employees (within 30 minutes)
- accessibility of a traffic cell
- number of pupils or training facilities
- sales
- prices
- training facilities

in order to mention only a few factors.

5.3 Traffic Flows and Estimation of the Parameters

It is plausible (and empirically testable) that the number of trips from a traffic cell i to other traffic cells j is proportional to the number of inhabitants of traffic cell i and to the corresponding individual travel probability. Furthermore, the individual travel rate depends on the attractivities of the traffic cells, by the resistance functions of the trips between the cells as well as on a mobility indicator. In total, the traffic flows $F_{ij}^{\alpha r}(t)$ of the

ITEM traffic model for a trip between the traffic cells i and j at time t with trip purpose α and trip mode r has the following functional form:

$$F_{ij}^{\alpha r}(t) = E_i(t)\varepsilon^{\alpha r}(t)b(t)g^{\alpha r}(t_{ij}^r)\exp(\gamma^{\alpha r}u_j^\alpha(\vec{E},\vec{x}) - u_i^\alpha(\vec{E},\vec{x})) \qquad (5.1)$$

Here, $E_i(t)$ is the number of population in cell i; $\varepsilon^{\alpha r}(t)$ is a scaling parameter which considers the flexibility of the actors to undertake a ride; $b(t)$ is the width of the hour groups; t_{ij}^r is the travel time depending on traffic density and trip mode; $g^{\alpha r}(t_{ij}^r)$ is the resistance function which depends on the travel time from cell i to cell j, parameter $\gamma^{\alpha r}$ describes the trips within the cell i. The $u_i^\alpha(\vec{E},\vec{x})$ are the attractivities of the particular traffic cells. The vector $\vec{x}$ represents the cell specific characteristics indicators of the infrastructure. All occurring values and parameters and their interdependencies are explained in detail explained in the appendix A1.

As experience shows one can rely on the assumption that actors compare the attractivities of the traffic cells. According to (5.1) this causes in case of the same resistance function with increasing attractivity difference $(u_j^\alpha(\vec{E},\vec{x}) - u_i^\alpha(\vec{E},\vec{x})) > 0$ an increase of the travel probability for a trip from cell i to cell j. Simultaneously, of course, the probability of travel in the opposite direction from cell j to cell i decreases. Since only the differences of the attractivities influence the transition rates, the sum of the attractivities over all traffic cells can be scaled to zero for each time interval τ:

$$\sum_i u_i^\alpha(\vec{E},\vec{x}) = \sum_i u_i^\alpha(\tau) = 0 \qquad (5.2)$$

For the characteristic resistance function $g^{\alpha r}(t_{ij}^r)$ the following ansatz has proved to be sufficiently flexible (HAUTZINGER, 1982; STEIERWALD/SCHOEN-HARTING, 1993a):

$$g^\alpha(t_{ij}) = (t_{ij}^r)^{b^{\alpha r}} \exp(-c^{\alpha r}t_{ij}^r) \qquad (5.3)$$

where, the parameters $b^{\alpha r}$ and $c^{\alpha r}$ are characteristic for the trip purpose and depend on the respective trip time interval and the traffic mode. In addition to the travel times t_{ij}^r between the traffic cells i and j the travel costs and corresponding comfort parameters or convenience parameters of travel from location i to destinations j by mode m, respectively, can be considered in the resistance function. In addition, there may be a fixed travel cost component as well as cost components taking account of network access at either end of a trip, waiting and transfer times at stations, waiting times at borders or congestion in metropolitan areas.

Within the Urban Area of Nanjing, the following travel purposes have to be distinguished:

- home – work
- home – education
- business
- visit
- shopping
- leisure
- hospital
- return

Simultaneously, the traffic mode groups “motorized individual traffic” and “public transport” are considered because changes in the modal split have to be assessed.

The parameters of the model (attractivities of the traffic cells, mobility indicator and resistance parameters) can now be determined from the empirical traffic flows and the travel time matrices by non-linear estimation methods. In order to achieve this goal we firstly minimize the deviations of the theoretical traffic flows, (i.e. of those flows which are calculated with the corresponding parameter values using the equation (5.1)), from the empirical traffic flows by means of the least square method and the corresponding adaptation of the parameters. The attractivities of the individual traffic cells, the mobility as well as the resistance parameters are directly determined in this first step. In a second step these calculated attractivities are set in relation to a bundle of influence factors via a multiple regression formalism. Thus the key-factors for the attractivities as well as their statistical weight are determined.

6. The Data Situation of Nanjing

The statistical significance of the results of the project depends on the data base. Therefore, we draw our attention in the next section to the quality and quantity of data.

In the years 1986/87 the last surveys in Nanjing referring to the 94 traffic cells were made. These surveys can be found in the "Data Situation" (see section 6.3 and 6.4). More recent data are available until 1995 and are mainly taken from the statistical yearbooks. But these data are divided into the 10 districts or even consist of all districts together. Other data than those listed in the "Data Situation" are either not available or too expensive, i.e. beyond the financial limits of the project.

6.1 The Traffic Network of Nanjing City for 1987 and 1996

The digitalized maps of Greater Nanjing as well as for Nanjing City are exhibited in the appendix A2. In the map of Greater Nanjing the 10 districts, 5 counties and 6 directions are depicted. The map of Nanjing City contains the subdivision of the 10 districts into 94 traffic cells plus 6 directions.

6.2 The Traffic Cells of Nanjing City plus 6 Directions

The digitalized maps of the traffic network of Nanjing City for 1987 and 1996 contain the headings:

- Roads
  - length of links
  - width of links
  - description and explanation of the six characteristics of streets (express artery, major artery, minor artery, minor road, highway and freeway), six control type of link ends/intersections (signalized, uncontrolled, roundabout, interchange, flared signal and priority), six control type of links (one way street, truck prohibit, bike prohibit, bus only, bicycla only, no management)

- Waterways
- Railways
- Bus lines

6.3 Socio-economic Data on the Level of the Traffic Cells

The data base according to the survey 1987, on the level of the 94 traffic cells plus the 6 directions used for the project looks as follows:

- Travel time matrix for the different traffic modes
- Geographical distance matrix
- O-D matrix (residents):
  - 8 trip purposes: home - work, home - education, business, visit, shopping, leisure, hospital, and return trips
  - 7 trip modes: bike, motorcycle, bus, by foot, O-Bus (Organization-Bus), taxi, car (private or organization), others (train, ship,...)
  - 4 trip time intervals (starting time): 5h - 10h, 10h - 15h, 15h - 20h, 20h - 5h
- Vehicle O-D matrix for residents:
  - 5 vehicles: small car, large car, truck, moto, others (police, fire, brigade, ambulance,...)
  - 7 purposes for passenger cars: work, education, travel, shopping, business, taxi, others
  - 6 purposes for trucks: goods transport empty, transport full, transport full - full (exchange of goods), shopping, return, others
- Rate of utilization
  - trucks
  - cars
- Average capacity
  - trucks
  - cars
- O-D matrix for immigration vehicle
- Traffic counts including map and hourly data for different vehicle types
- Maps plus description files (containing more detailed information concerning the floor space in the CBD (4 traffic cells)) of:
  - existing land use in 1990
  - planned land use of 2010
- Population survey
  - number of population

6.4 Socio-economic Data on the Level of the 10 Districts

The data base on the level of the ten districts used for the project is listed below:

1. Yearly data base between 1987 and 1995:

➢ Population numbers

2. Data for the years 1987 and 1995:

➢ Work places according to professions
    - agriculture
    - industry
    - construction
    - transportation and communication
    - services (total, public, hotel)
    - culture
    - education
    - science
    - sports
    - hospital
    - administration

➢ Universities, colleges
    - number of students

6.5 Global Socio-economic Data of Nanjing City

The data base on the level of Nanjing City is listed below:

1. For 1987:

➢ Floor space per capita for the urban area

2. For the years 1987 - 1996:

➢ Number of vehicles:
    - bikes
    - busses
    - cars owned by private persons

  - cars owned by firms
  - trucks
- Travelling costs for different years for public traffic (bus, train, ship. taxi)
- Household budget

7. Adaptation of ITEM to the Specific Data Situation of Nanjing

In the case of Nanjing any mathematical model must be adapted to the specific data situation of the configuration space. All parameters and variables of the model have to be specific to the sectors in which data exist in such a way that the model is able to describe the historical situation. With these results one obtains a calibration of the model for the period 1986 - 1996. Taking the calibrated model, forecast scenarios can thereupon be set up. In those sectors where no yearly data exist, one has to modify the model by making plausible assumptions.

However, one has to modify the very detailed model described above with some aspects considered in the next two subsections.

7.1 Adaptation of the Traffic Sector - Consequences

The last big survey referring to the traffic sector on the level of the 94 traffic cells plus the six directions was made 1987. Since thiis thorough survey dates back more than one decade, some problems and some modelling consequences arise which will now be discussed:

Firstly, in the O-D-matrices more than 90% of all entries are zero. The O-D-matrices are origin-destination-matrices which describe the detailed traffic flows from each of the 94 traffic cells to each other with respect to a special trip purpose, a special trip mode, a special trip time interval and the corresponding time. It is for instance clear that nearly all entries referring to the trip mode "private car" must vanish because in 1987 only 85 private cars have been registered in Nanjing. Considering the 94 times 94 = 8836 entries of one O-D-matrix - which belongs to one trip purpose and one trip time interval - it is not surprising that almost all O-D-matrices have no entries also for motorcycles, organization buses, taxis and other traffic modes like train or ship. In order to use only those O-D-matrices with at least 20% entries one has to restrict oneself to the following *main trip modes, trip purposes* and *trip time intervals:*

Traffic mode	Trip purpose	Trip time interval
Bike	Home-work	5h-10h
Bus	Home-education	10h-15h
By foot	Shopping	15h-20h
	Return	

Therefore only $3*4*3=36$ O-D-matrices have been investigated in detail. The traffic mode "car" which will become relevant in future must be treated in a different way. Here one has to take into account several aspects concerning the actual traffic situation of Nanjing as well as the already existing planning measures in the transportation sector and in the urban and regional sector.

Secondly, the distribution of the traffic modes and with them the appearance of the roads was totally different in 1987. Therefore, comparing different modes from the survey 1987 with the traffic situation today contains some difficulties. The consequence is the same as in the above mentioned case: Only those modes will be considered in detail which have enough entries. The comparison with all other traffic modes follows by taking into account the global data and the knowledge of the actual traffic situation.

Thirdly, with an increasing traffic density the travel time matrices have certainly also changed during the last ten years. Here, it will be helpful not to use the travel time matrices from the survey 1987 but calibrated values which include the effect of the distances between the traffic cells. A linear dependence between travel time and distance may be useful as long as no landscape obstacles exist on the route. In this case the travel times can be estimated with the help of the digitalized maps as well as with the existing travel time matrices from 1987 which may be a good control for the relative distances between the 94 traffic cells.

Next, with respect to many further details like travel time matrices, rates of utilization, average capacities, O-D-matrices for immigration vehicles etc. there do not exist enough data or only global data are available. With respect to these aspects it is at least possible to make alternative assumptions and to calculate the corresponding scenarios on the basis of the model. Especially the planning measures of the CBD with respect to the separation or the mixing of land use should be mentioned in this context. Different planning measures lead to different amounts of induced traffic which can be estimated by this scenario technique.

7.2 Adaptation of the Urban/Regional Sector - Consequences

With respect to migration the main problem in the urban and the regional sector is also the shortage of the data base. In detail the data situation is as follows:
All population numbers only exist on the level of the ten districts except for the data of the population survey 1987 which was made on the level of the 94 traffic cells. How the problem "Determination of regional utilities and mobilities for migrating population" was solved can be seen below in this section.

Especially in the urban and regional sector far-reaching measures with a great influence on the migration and the housing situation are planned. It is worth mentioning, that maps plus description files which contain more detailed information concerning the floor space in the CBD for the existing land use in 1990 as well as the planned land use of 2010 may be helpful. But firstly these maps in addition to the population numbers on the district-level are too coarse-grained data for a differentiated forecasting. Secondly, it is difficult to keep long term planning measures in an updated state with respect to the rapidly developing situation of China or Nanjing. Nevertheless, such planning measures have a trendsetting character. Such trends can again be portrayed by scenario techniques.

Determination of Regional Utilities and Mobilities for Migrating Population:

We consider L cells with the time-dependent population numbers $E_1(t),...,E_L(t)$, the total population is $E(t)$. Firstly, we introduce the relative variables x_i (fraction share of total population):

$$E_i(t) = E(t) \cdot x_i(t) \tag{7.1}$$

The x_i fulfil the norm condition

$$\sum_{i=1}^{L} x_i(t) = 1 \tag{7.2}$$

Using further the relation

$$x_i(t) = \exp(2V_i(t)) \cdot C(t) \tag{7.3}$$

we obtain $(L+1)$ variables to be determined: the utilities $V_1,\ldots,V_L$ and C. With the normalization condition (7.2) it follows for $C(t)$:

$$C(t) = \left(\sum_{i=1}^{L} \exp(2V_i(t))\right)^{-1} \tag{7.4}$$

Since, on the other hand, there are only L empirical variables, we demand the additional constraint:

$$\sum_{i=1}^{L} V_i(t) = 0 \tag{7.5}$$

This yields to

$$\prod_{i=1}^{L} x_i(t) = C^L(t) \cdot \exp\left(2\sum_{i=1}^{L} V_i(t)\right) = C^L(t) \tag{7.6}$$

and therefore

$$C(t) = \left(\prod_{i=1}^{L} x_i(t)\right)^{1/L} \tag{7.7}$$

that means (7.4) as well as (7.7) are fulfilled as long as the additional constraint (7.5) is fulfilled. Now, we introduce the evolution utility $W_i(t+1,t)$, which depends on two successive time steps (usually years) and on the utilities at these times $(V_i(t), V_i(t+1))$:

$$\frac{x_i(t+1)}{C(t+1)} = \exp(2W_i(t+1,t)) \cdot \frac{x_i(t)}{C(t)} \tag{7.8}$$

With (7.3) we receive the relation between the utilities:

$$V_i(t+1) = W_i(t+1,t) + V_i(t) \tag{7.9}$$

Since the condition for the V_i (7.4) has to be fulfilled, we obtain

$$\sum_{i=1}^{L} W_i(t+1,t) = 0 \tag{7.10}$$

The $V_i(t)$ are measures of the attractivity for migrants of cell i at time t. The quantities $W_i(t+1,t)$ are measures for the – positive or negative – change of this attractivity in the time interval [t,t+1].
Let us now consider the equations of migration:

$$\begin{aligned}\Delta_i^e(t+1,t) &\equiv x_i(t+1) - x_i(t)\\ &= \sum_{j=1}^{L}\left(p_{ij}(t+1,t)x_j(t) - p_{ji}(t+1,t)x_i(t)\right)\end{aligned} \tag{7.11}$$

with

$$p_{ji}(t+1,t) = \nu(t) f_{ji} \exp\left(V_j(t+1) - V_i(t)\right) \tag{7.12}$$

then it follows from (7.9)

$$\Delta_i^e(t+1,t) = \sum_{j=1}^{L} \nu(t) f_{ji} C(t) \exp\left(V_i(t) + V_j(t)\right)\cdot\left(\exp\left(W_i(t+1,t)\right) - \exp\left(W_j(t+1,t)\right)\right) \tag{7.13}$$

We assume $f_{ij} = f_{ji}$ to be given:
Then we obtain the following $L \cdot T$ equations

$$\Delta_i^e(t+1,t) = \nu(t)\cdot D_i(t+1,t)\,;\, i = 1,\ldots,L\,;\, t = 0,\ldots,T-1 \tag{7.14}$$

with the abbreviation

$$D_i(t+1,t) \equiv \sum_{j=1}^{L} f_{ji} C(t) \exp\left(V_i(t) + V_j(t)\right)\cdot\left(\exp\left(W_i(t+1,t)\right) - \exp\left(W_j(t+1,t)\right)\right) \tag{7.15}$$

It is true that these $L \cdot T$ equations will in general not be fulfilled exactly but by fitting of the mobilities $\nu(t)$ (T variables for T time steps) the following expression can be minimized:

$$\sum_{i=1}^{L}\left(\Delta_i^e(t+1,t)-\nu(i)D_i(t+1,t)\right)^2 = Min. \tag{7.16}$$

Differentiation with respect to $\nu(t)$ leads to

$$0=\sum_{i=1}^{L}2\left(\Delta_i^e(t+1,t)-\nu(t)D_i(t+1,t)\right)\cdot D_i(t+1,t) \tag{7.17}$$

and therefore

$$\nu(t)=\frac{\overrightarrow{\Delta^e}(t+1,t)\cdot\overrightarrow{D}(t+1,t)}{\overrightarrow{D}(t+1,t)\cdot\overrightarrow{D}(t+1,t)} \tag{7.18}$$

7.3 Summary of the Algorithms – Analysis and Forecasting

Since essential data about population and migration in Nanjing are lacking, the migration can only be calculated indirectly with the help of methods described in subsection 7.2.

Empirical data of the 94 traffic cells with respect to several sectors are only available for the years 1986/87. Hence, we have to practice the so-called scenario technique. Especially, the following steps can be visualized:

A) Basic scenario

1. Updating of the numbers of population and workplaces for the 94 traffic cells with respect to the numbers of population and workplaces for the 10 districts from 1987 to 1995.
2. Forecast of the numbers of population and workplaces for the 10 districts with respect to the planning as defined in the area utilization plan; estimation of the migration flows from outside and between the 10 districts.
3. Forecast of the numbers of population and workplaces for the 94 traffic cells by means of the forecast for the 10 districts.
4. Updating of the vehicle stock.
5. Division and updating of the directions into
   a) through traffic
   b) source traffic.

6. Traffic generation, allocation, division and apportion for the years 1987, 1995, 2000, 2005 and 2010 for the basic scenario.

B) *Alternative scenario* for the year 2000: mixed land use on the district level on the level of the 94 traffic cells.

C) Small sized mixed land use for the year 2000. Relation between employed persons and workplaces 1:1 on the level of the 94 traffic cells.

D) Separation of landuses on the district level: Relation between employed persons and workplaces not 1:1 on the level of the 94 traffic cells.

E) Comparative analysis and consequences for the being within reach of the districts in the different scenarios.

By means of these scenarios we will be able to give recommendations about the installation of the planned undergrounds, about

- the further enlargement of the street network within Nanjing
- the further enlargement of the street network into the 6 directions,
- the situation of the bikers,
- the consequences of the separation of land use / mixed land use
- etc.

Part III

Presentation of the Scenarios, Results of the Calculations, Conclusions and Recommendations

Part III

8. Methodology (Analysis and Forecasting) of the Traffic and Urban/Regional Situation of Nanjing City

The spatial sciences are still far away from a comprehensive theory of urban development.

In the spatial sciences there exist a large variety of observations and partial hypotheses about individual phases and aspects of urban development. With some risk of generalisation, four directions of urban development theory can be distinguished: technical, economic, social and political theories (Wegener, 1998b). Consider the first two directions:

- *Technical* theories consider cities primarily as mobility systems. Technical conditions determine the internal organisation of cities. The high density of the medieval city resulted from the need for fortifications and from the fact that most trips had to be made on foot. When these two constraints disappeared in the 19th century, urban development, following this paradigm, largely became a function of transport technology. After the introduction of railways cities expanded on both sides along the railway lines fanning out from the city centre. With the diffusion of the private automobile, first in America and after World War II also in Europe, the areas between the railway lines could also be used for housing, and so the expansion of urban areas became less directed and more dispersed. Thinking in this paradigm the development of the urban area of Nanjing is among other factors determined by two opposite effects: First, the resulting concentration effect by the traffic modes foot and bicycle and second, by the tendency to expand along the six main transport directions.

- *Economic* theories see cities essentially as a system of markets on which supply and demand interact: the urban labour market, the urban housing market and the urban land market. A fundamental assumption of all spatial economic theories is that locations with good accessibility are more attractive and have a higher market value than peripheral locations. This fundamental assumption goes back to von Thünen (1826) and has since been varied and refined in many ways. For urban analysis the most influential refinement is the model of the urban land market by Alonso (1964). The basic assumption of the Alonso model is that firms and households choose that location at which their bid rent, i.e. the land price they are willing to pay, equals the asking rent of the landlord, i.e. that the land market is in equilibrium. The high growth rates (8% per year) of the region of Nanjing underlines its economic market function. In a broader concept the region of Nanjing has to be seen in competition with other cities like Shanghai among population, workplaces, capital and political influence.

The relationship between urban land use and transport can similarly be seen as a self-reinforcing positive feedback loop in which urban development generates traffic and, - vice versa, - transport opportunities stimulate urban development. Congestion and land prices, rents and accessibility indicators, to mention an few factors of influence, act as equilibrating negative feedback factors.

For the development of ITEM, an idealised urban transport model is sketched out as a benchmark. Eight types of major urban subsystem are distinguished (see also EUROSIL, WP 7, 1998). They are categorised by their speed of change:

- *Very slow change: networks, land use.* Urban transport, communications and utility *networks* are the most permanent elements of the physical structure of cities. Large infrastructure projects require a decade or more, and once in place, are rarely abandoned. The *land use* distribution is also rather stable.

- *Slow changes: workplaces, housing.* Business buildings have a life-span of up to one hundred years and take several years from planning to completion. *Workplaces* (non-residential buildings) such as factories, warehouses, shopping centres or offices, theatres or universities exist much longer than the firms or institutions that occupy them, just as *housing* exists longer than the households that live in it.

- *Fast change: employment, population.* Firms are established or closed down, expanded or relocated; this creates new jobs or makes workers redundant and so affects *employment.* Households are created, grow or decline and eventually are dissolved, and in each stage in their life-cycle adjust their housing consumption and location to their changing needs; this determines the distribution of *population.*

- *Immediate change: goods transport, travel.* The location of human activities in space gives rise to a demand for spatial interaction in the form of *goods transport* or *travel.* These interactions are the most volatile phenomena of spatial urban development; they adjust in minutes or hours to changes in congestion or fluctuations in demand.

There is a ninth subsystem, the *urban environment.* Its temporal behaviour is more complex. The direct impact of human activities, such as transport noise and air pollution are immediate; other effects such as water or soil contamination build up incrementally over time, and still others such as long-term climate effects are so slow that they are hardly observable. All other eight subsystems affect the environment by energy and space consumption, air pollution and noise emission, whereas only locational choices of housing investors and households, firms and workers are co-determined by environmental quality, or lack of it. All nine subsystems are partly market-driven and partly subject to policy regulation.

In this section the methodology of analysing and forecasting the development of the transport system and the regional structure of Nanjing is described in some detail. The ITEM transport model is used to analyse in a first step the transport system of Nanjing in 1987 (Section 8.1). In a second step the further development of the transport system is simulated for different scenarios (basic scenario E and scenarios A ,B, C, see section 10). Not only the traffic between the cells, but also the traffic within the cells is modelled (see section 8.2.). Moreover the development of the regional structure in terms of the evolution of the population distribution has to be considered. For this aim a Multinomial Logit Model (MNL) is used in section 8.3. The methodology in general is described in section 8.4. Here also the interactions between regional structure and transport system are considered.

8.1 The ITEM Transport Model

The traffic flows for the ITEM transport model from cell i to cell j with mode r are given by the following functional form (see (5.1) with (5.3))

$$F_{ij}^{\alpha r}(t) = E_i(t)\varepsilon^{\alpha r}(t)(t_{ij}^r)^{b^{\alpha r}} \exp(-c^{\alpha r} t_{ij}^r) \exp(\gamma^{\alpha r} u_j^{\alpha r}(\vec{E},\vec{x}) - u_i^{\alpha r}(\vec{E},\vec{x})), \quad (8.1)$$

with

$$\gamma^{\alpha r} = \delta_{ij}\gamma_1^{\alpha r} + (1-\delta_{ij})\gamma_2^{\alpha r},$$

where

$$\delta_{ij} = \begin{cases} 1, \mathit{for}\ i = j \\ 0, \mathit{for}\ i \neq j \end{cases}$$

is the Kronecker symbol. $E_i(t)$ is the population number of cell *i* and t_{ij}^r is the travel time from cell *i* to *j* by mode *r* and trip purpose α. Further parameters of the model are:

- attractivities $u_i^{\alpha r}$ of the cells *i*, that depend on characteristics of the cells x_i^n like population number or number of workplaces, accessibility measures to embrace a few location parameters, trip purpose and so on:

$$u_i^{\alpha r} = \sum_i b_i^{n\alpha r} \cdot x_i^{n\alpha r}$$

  It depends on trip purpose and travel mode which variables are significant for the attractivities.
- parameter $\gamma^{\alpha r}$ to describe trips within the cell *i*, split into $\gamma_1^{\alpha r}$ describing the diagonal elements of the matrix because of $\gamma^{\alpha r} = \gamma_1^{\alpha r}$ for $i = j$ and $\gamma_2^{\alpha r}$ describing the non-diagonal elements of the matrix because of $\gamma^{\alpha r} = \gamma_2^{\alpha r}$ for $i \neq j$,
- scaling parameter (mobility parameter) $\varepsilon^{\alpha r}$ for the modes *r* and
- the parameters of the resistance function $b^{\alpha r}$ and $c^{\alpha r}$

The ITEM transport model will be used in the following to analyse first the transport system in Nanjing in 1987. In a second step the further development of the transport system will be simulated for a select group of scenarios. Moreover the development of the regional structure like the population distribution has to be considered. For this a MNL-model is used, that means the population number E_i of the cell *i* is given by

$$E_i = E_{tot}(t) \cdot \frac{\exp(2\widetilde{V}_i)}{\sum_i \exp(2\widetilde{V}_i)}$$

with the total population number E_{tot} and attractivities $\widetilde{V}_i$. These "population attractivities" depend on characteristics of the cells like number of workplaces which are used as independent variables. Also variables like accessibilities or even transport attractivities for different modes could be significant for the population attractivities of the cells. Therefore the interaction with the transport system can be described.

8.2 Calculations of the Inner-cell Travel Times t_{ii}

Since in all scenarios the classical traffic assignment procedures are not able to compute adequate travel times t_{ii} within one cell i, we have to develop an appropriate inner-cell travel time procedure. The following procedure will show a method to calculate the diagonal elements t_{ii} of the travel time matrix (t_{ij}) where the non-diagonal elements t_{ij} $(i \neq j)$ are given as well as the distance matrix (d_{ij}) between the traffic cells *i* and *j*, the original travel time matrix (t_{ij}^o) from the survey of 1986 and a digitalized map with all traffic cells:

First, we introduce $\widetilde{t}_{ik}$ as the arithmetic average of t_{ik} and t_{ki} which may differ so that we obtain

$$\widetilde{t}_{ik} = \frac{t_{ik} + t_{ki}}{2}$$

Next, we assume for neighbouring traffic cells k the proportionality

$$\frac{d_{ik}}{\widetilde{t}_{ik}} \cong \frac{d_{ii}}{t_{ii}},$$

where d_{ii} is an average trip distance in the cell i which still has to be determined.

Now let N_i be the set $\{i_1, i_2, ...\}$ of direct neighbour cells of cell i and let n_i be the number of elements of N_i. Then it follows in a very natural way for t_{ii}:

$$t_{ii} = \frac{1}{n_i} \sum_{j \in N_i} \tilde{t}_{ij} \cdot \frac{d_{ii}}{d_{ij}} \tag{8.2}$$

For the determination of d_{ii} it is important to know that the average distance of two coincidentally chosen points on the area of a circle with radius r is $0{,}545 \cdot r$. For a cell with quadratic base (base area $4a^2$) the average distance of two points is $0{,}385 \cdot a$. Since we cannot assume that the streets are straight, we choose $d_{ii} = 0{,}5 \cdot a$. Here, the cell is assumed to be quadratic which seems to be a good approximation (see the map of Nanjing-Urban Area). This choice leads to

$$d_{ii} = \sqrt{\frac{A_i}{4}} \cdot 0{,}5 = \sqrt{\frac{A_i}{8}} \tag{8.3}$$

where A_i is the known area of cell i. Equation (8.3) with (8.2) yields a recipe to calculate the inner-cell travel times t_{ii}. Insertion of t_{ii} completes the travel time matrices.

8.3 Procedure in General

In the figures 8.1 to 8.3 at the end of this subsection the framework of the ITEM transport model is illustrated for the region of Nanjing. In the following eight steps we summarize the procedure in schematic terms:

Step 1 - Traffic assignment

In the first step the traffic flows $F_{ij}^{car}(1987)$ for the considered transport modes are assigned on the corresponding network (1987). From this the travel times $t_{ij}^{r}(1987)$ between the single cells as well as accessibility measures of the cells for the considered modes are determined.

In the same way the travel time matrices t_{ij}^{r} *(iteration 1)* and the corresponding accessibility measures are determined for the traffic flows $F_{ij}^{\alpha r}$ (1987) and the network for the year 2000 for the planning scenarios.

Step 2 - Estimation of the parameters

From the traffic flows $F_{ij}^{\alpha r}(1987)$ for the considered modes, from the corresponding travel times $t_{ij}^{r}(1987)$ and from the population numbers $E_i(1987)$ the parameters of the ITEM transport model are estimated by a non-linear estimation procedure:

- attractivities $u^{\alpha r}(1987)$ of the cells *i*,
- parameters $\gamma^{\alpha r}$,
- scaling parameters (mobility parameters) ε^{r} for the modes *r* and
- the parameters of the resistance function b^{r} and c^{r} .

Statistical tests (correlation, Fisher's F-value) are used to evaluate the quality of the parameter estimation.

Step 3 - Attractivity parameters (multiple regression)

By means of a multiple regression the dependence of the attractivities of the cells on the characteristics of the cells (for example population and employee numbers, accessibilities) is determined. That means the corresponding parameters b_i^{n} and β_i^{n} (influence of the independent variables, elasticities) as well as statistical test values are calculated.

Step 4 - Analysis of the population distribution

To analyse the population distribution and its dependence on the accessibilities of the cells the "population attractivities" (MNL model) are calculated from the population distribution 1990. The influence of the changed accessibility measures of the cells for the considered planning scenarios on the "population attractivities" can be determined by a multiple regression using accessibility measures for a basic scenario.

In the fifth step an **iteration procedure** (step 5 to 7) to forecast the evolution of the transport system and the population distribution for the planning scenarios is employed.

Step 5 - Simulation of the population development

If it is known which accessibility measures are significant for the population attractivities, in the next step these attractivities and therefore the changed population distribution E_i (*iteration 1*) can be calculated for the planning scenarios till the year 2010. Furthermore changed migration flows into or out of the considered region can be taken into account.

Step 6 - Simulation of the traffic flows

In this step the (transport) attractivities $u^{\alpha r}$ *(iteration 1)* of the cells, that can also depend on the population distribution and on accessibilities, can be calculated for the different considered planning scenarios as well.

With these attractivities u_i *(iteration 1)*, with the travel times t_{ij}^r (*iteration 1*) and the population numbers E_i (*iteration 1*) of step 5, the traffic flows $F_{ij}^{\alpha r}$ (*iteration 1*) are determined for both modes r. The mobility parameter and the parameters of the resistance function were assumed to be constant.

Step 7 - Assignment of the traffic flows

The traffic flows $F_{ij}^{\alpha r}$ (*iteration 1*) are now assigned for each planning scenario on the corresponding network to get the new travel times t_{ij}^r (*iteration 2*) and accessibility measures. With these travel times and accessibility measures for each mode the population configuration (dependent on the accessibilities) and the traffic flows change again, so that an iteration procedure has to be applied.

To accelerate this iteration procedure the average between the new travel times t_{ij}^r (*iteration 2*) and the travel times t_{ij}^r (*iteration 1)* of the step before were taken to calculate the next step of the procedure. Step 5 to 7 are repeated till the change of the flows and travel times can be ignored, in other words when the procedure converges. As result one gets the traffic flows $F_{ij}^{\alpha r}$ for the single modes r and trip purposes α for the different planning scenarios until 2010.

Step 8 - End of the iteration procedure

The comparison of the traffic flows $F_{ij}^{\alpha r}$ for the planning scenarios with the basic scenario shows the effects of the changed modal split and the interaction with the population development. Even direct and indirect induced traffic shares are considered by the model.

From the traffic flows $F_{ij}^{\alpha r}$ and their assignment on the corresponding network the traffic performances and so on can be determined for the single schemes.

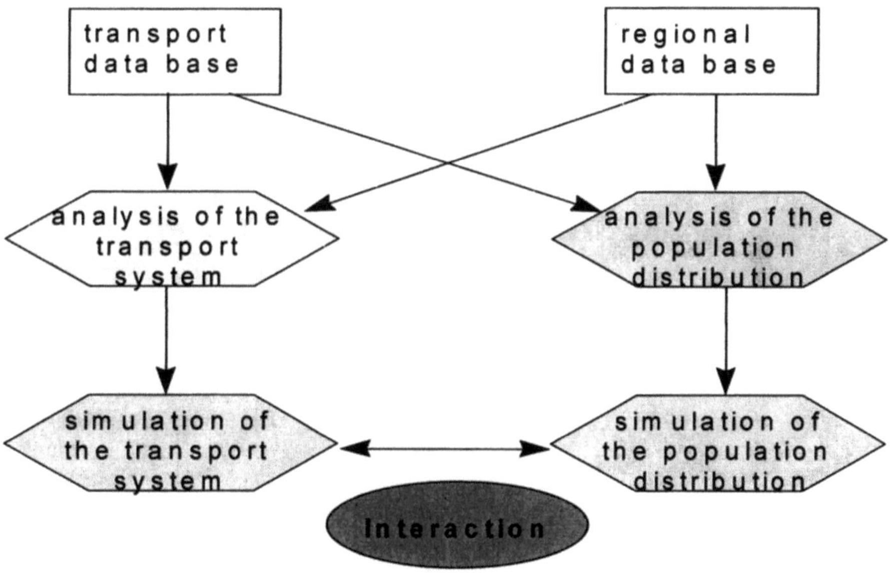

Figure 8.1: Procedure in general

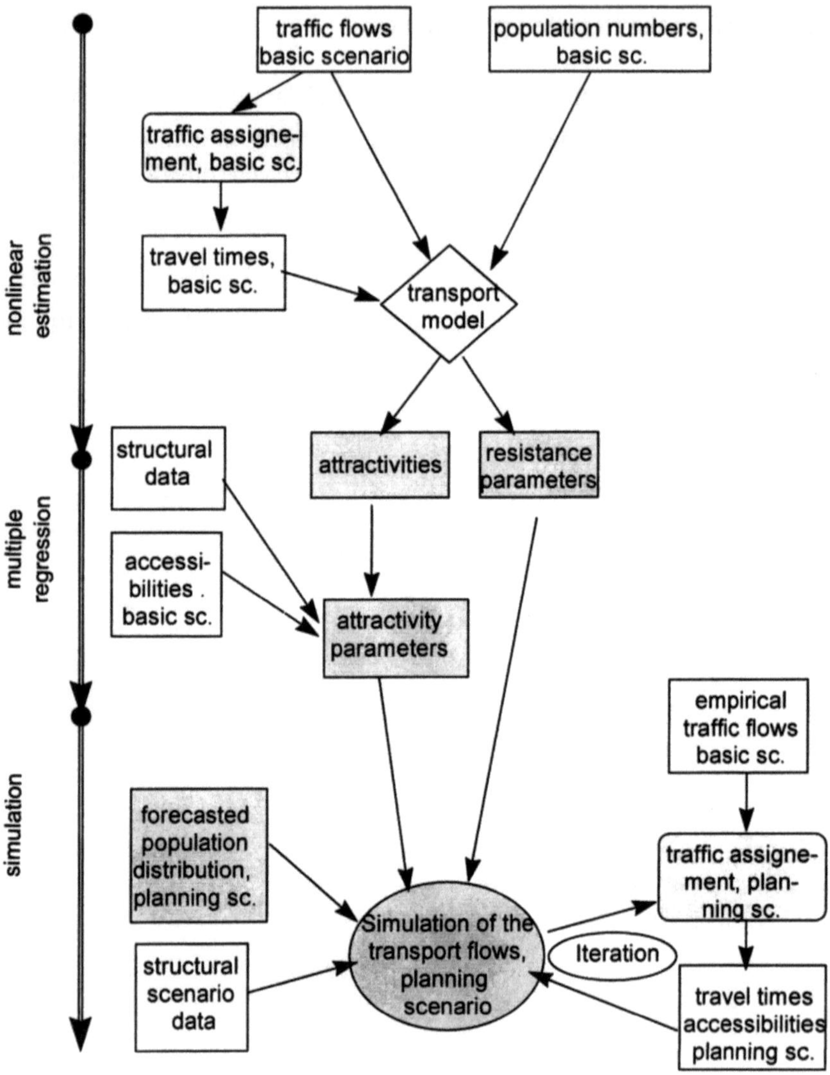

Figure 8.2: Detailed Procedure Concerning the Dynamic Transport Model

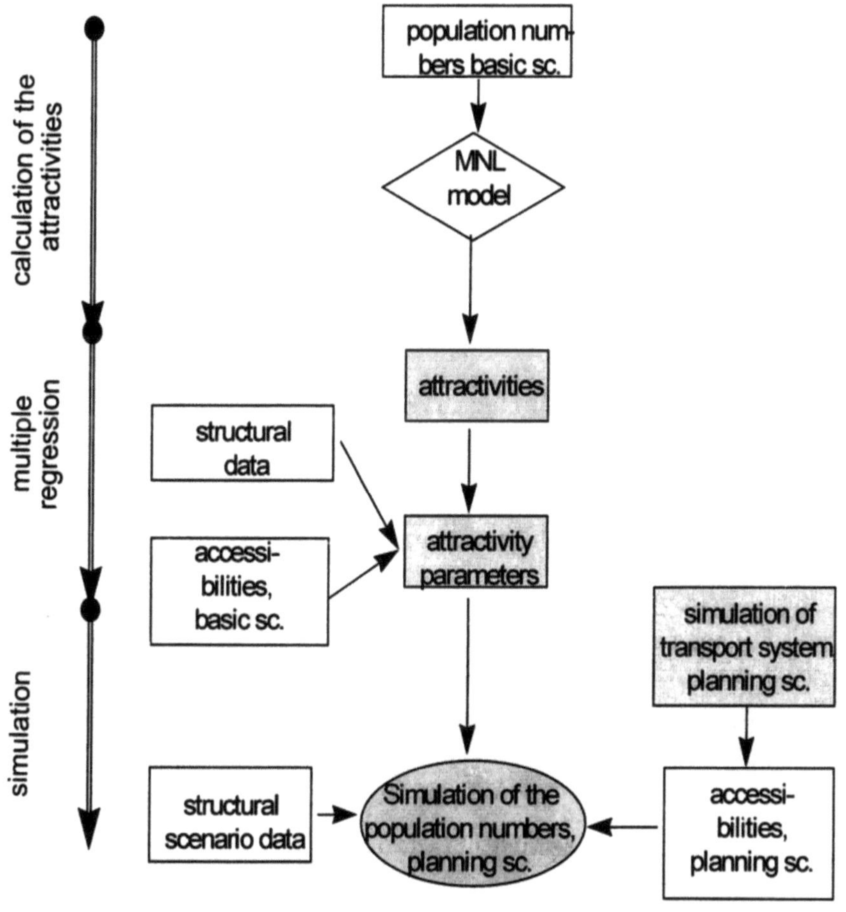

Figure 8.3: Detailed Procedure Concerning the Population Development

9. Analysis of the Transport System and of the Population Development

This chapter gives an overview of the results of the analysis in the transportation and the urban/regional sector.

In the transportation sector the estimation of the occurring parameters in the formula (8.1), which form the resistance functions, are listed in dependence upon the trip modes and trip purposes in table 9.1. The resistance functions are plotted in fig. 9.1. In addition the best results of the multiple regression formalism are presented in table 9.2, again depending on different trip modes and trip purposes. This table contains the most significant variables of the appropriate attractivities. The attractivities with respect to the traffic are also depicted in the figures A3.1a, A3.1b, and A3.1c which can be found in the appendix A3.

In the urban/regional sector the existing, respectively calculated, population numbers for the years 1987 until 2010 are depicted for the three selected years 1987, 1996 and 2010 in the figures A3.2a, A3.2b, and A3.2c in the appendix A3. Table 9.3 shows the best multiple regression result for the appropriate attractivity.

9.1 Results in the Transportation Sector

As described in chapter 8, in a first step the parameters of the ITEM transport and evolution model are estimated from the empirical traffic flows $F_{ij}^{\alpha r}$, the corresponding travel time matrices t_{ij}^{r} and the population numbers E_i of the cells i. The results are

- attractivities $u_i^{\alpha r}$ of the cells i,
- scaling parameter (mobility parameter) $\varepsilon^{\alpha r}$ for the modes r and
- the parameters of the resistance function $b^{\alpha r}$ and $c^{\alpha r}$.

In the following step the attractivities of the cells are investigated by a multiple regression procedure concerning their dependence upon the characteristics of the cells. The results of this analysis is presented in the following and is the basis for the simulation of the different scenarios (section 10), which have to be considered.

Estimation of the Parameters

Table 9.1 presents the estimation of the parameters except attractivities for the travel modes **bike, bus** and **by foot** and for the trip purposes **home-work, home-education, shopping** and **return** in the first time interval (5h – 10h).

Table 9.1: Estimation of the Parameters for the first time interval (5h – 10h)

	ε	b	c	γ	F	R^2
Purpose: homework						
Bike	7.4 E-04	0.071	0.179	15.799	77	0.467
Bus	8.2 E-04	2.5 E-04	0.029	0.201	9	0.107
By foot	1.1 E-04	0.056	0.224	-1.363	277	0.761
Purpose: home-education						
Bike	6.9 E-04	3.3 E-04	0.100	1.509	76	0.495
Bus	8.0 E-04	0.010	0.054	0.100	7	0.157
By foot	2.6 E-03	0.524	0.137	-1.310	2722	0.968
Purpose: shopping						
Bike	6.6 E-04	9.6 E-03	0.084	1.943	53	0.388
Bus	7.7 E-04	1.4 E-05	0.092	1.285	143	0.626
By foot	9.8 E-04	1.192	0.182	-1.189	189	0.680
Purpose: return						
Bike	6.7 E-04	0.396	0.186	-0.740	51	0.384
Bus	4.5 E-04	4.6 E-05	0.019	0.190	30	0.312
By foot	1.0 E-03	0.736	0.197	-1.475	183	0.674

The results for the mobility parameter (scaling parameter ε) which can be seen as a measure for the mobility of the agents can be found in the first column of the table. The mobility parameter can only be seen in the context of the respective resistance function.

The next two columns of table 9.1 show the results for the estimated parameters b and c of the resistance function (travel time in minutes). Here are big differences between the different traffic modes and the different trip purposes. To illustrate these differences in figure 9.1 the resistance functions multiplied by the mobility parameter of all three modes and all four trip purposes are depicted in their dependencies on the travel times.

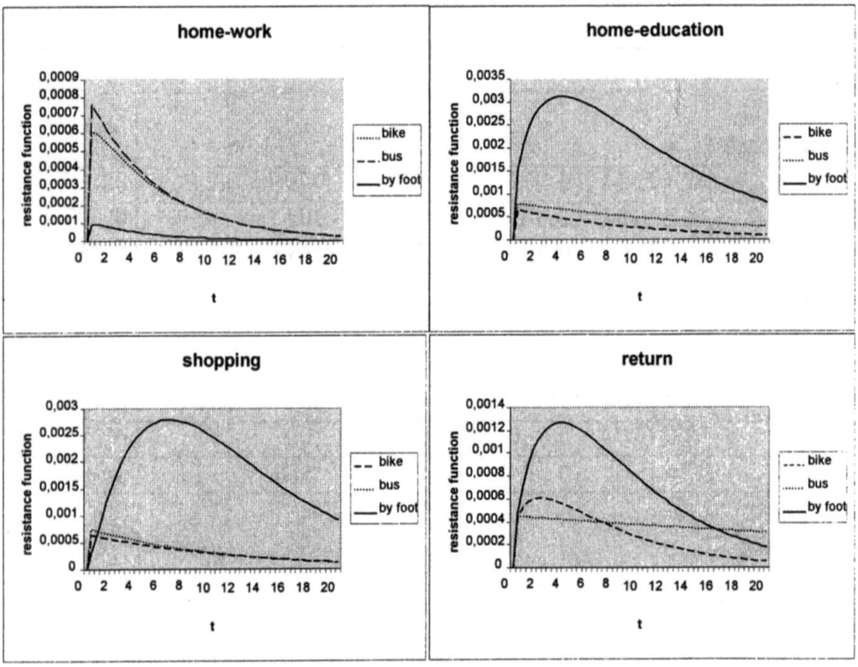

Figure 9.1 Resistance Functions for the different trip purposes home-work, home-education, shopping and return and for the trip modes bike, bus and by foot for the first time interval (5h – 10h)

All resistance functions of the figure 9.1 show the typical behaviour with a single maximum at more or less small trip times and a decrease for $t \rightarrow 0$ towards the origin (0/0) and a decrease for $t \rightarrow \infty$ against the zero-axis. The most preferred trip mode for the trips from home to work is the bus followed by the bike whereas by far most Chinese prefer walking to their place of education in Nanjing. Another convincing aspect is that the people go shopping by foot with a maximum of the resistance function of more than the threefold of the appropriate function for bike and bus. The trip modes with purpose return show a similar behaviour but with smaller values because the explored time interval is from 5h until 10h.

The parameter γ seems to have a very unforeseeable behaviour in dependence upon the modes and the purposes. That means that the role which the inner-cell flows play is differing very much with respect to the modes and purposes. Remember here the special shape of $\gamma^{\alpha r}$

$$\gamma^{\alpha r} = \delta_{ij}\gamma_1^{\alpha r} + (1-\delta_{ij})\gamma_2^{\alpha r}.$$

Therefore a negative entry for γ in table 9.1 normally means a positive γ_1 (trip within a cell) and a negative γ_2 (trip between different cells) which is quite convincing for the trip mode foot.

In table 9.1 the correlation and the Fisher's F-value for the comparison of the given ("empirical") flow matrix with the "modelled" matrix, calculated by the ITEM transport model with the estimated parameters, are presented in the last two columns. These values can be seen as measures, that indicates the correspondence between the original data base and the estimated values and therefore the quality of the results of the estimation.

The result for the estimation in this case study show a high significance with a correlation coefficient of more than 0.6 for the five cases:

- trip purpose: home-work mode: by foot
- trip purpose: home-education mode: by foot
- trip purpose: shopping mode: bus
- trip purpose: shopping mode: by foot
- trip purpose: return mode: by foot

Attractivities

The transport attractivities on the level of the 94 Urban Area are shown in figures A3.1a, A3.1b, and A3.1c in the appendix A3. For the trip purposes home-work, home-education and shopping for the year 1987 and again for the first trip time interval (5h – 10h).
While comparing the three figures A3.1a, A3.1b, and A3.1c one recognizes that the attractivities for home-work seem to be nearly equal distributed over the whole Urban Area which indicates the practising of

mixed landuse in Nanjing, 1987. 88 of the 94 cells have a positive attractivity and since the sum of all attractivities is zero the 6 remaining cells must be very disattractive. In contrast, the attractivities for the purposes shopping and home-education show a more differentiated behaviour. That means that shopping centers on the one hand and schools, universities, etc. on the other hand are more localized and not as equally distributed as working places.

Dependency of the Attractivities from the characteristics of the cells

The attractivities - estimated in the first step - are in the following related to the characteristics of the traffic cells such as population numbers or accessibilities by a multiple regression. On the base of the 94 traffic cells the following (independent) variables are used:

- population numbers
- average speed bike
- average speed bus
- average speed foot
- dwelling area
- business area
- public area
- industrial area
- greenbelts
- municipal area (gas works, water works, ...)
- storage area (storehouses)
- especially area (public leisure centers)
- external traffic area (railway stations, docks, etc.)

These variables were scaled to zero - corresponding to the attractivities ("z-transformation", see annex).

The results are summarized in the tables 9.2, again for the three traffic modes **bike, bus and by foot** and for the four trip purposes **home-work, home-education, shopping** and **return**. Only the best results, that means those results with the highest significance, are shown.

One line of the tables corresponds to one regression result. The parameter β gives the relative influence of the respective variable on the attractivities. Furthermore the significance of each variable is given in parenthesis (*Sig.*

T). In the last column the statistical test values (R, R^2, Fisher's F-value) can be found.

Tables 9.2: Regression results for the transport attractivities

1. Purpose: home – work; mode: bike

	average speed: bus	**industry**	R R^2 F**-value**
β **(Sig. T)**	-0.329 (0.003)	0.233 (0.035)	0.315 0.099 5.027

In words: The attractivity in this case will be formulated as a function of the parameters industrial areas (positive influence) and the average speed over all cells by bus (negative influence).

2. Purpose: home – work; mode: bus

	Population	**business**	R R^2 F**-value**
β **(Sig. T)**	0.437 (0.000)	0.266 (0.005)	0.597 0.356 25.144

In words: The attractivity in this case will be formulated as a function of the parameters population and the business areas (both with positive influence).

3. Purpose: home – work; mode: by foot

	Business	**greenbelt**	**municipal**	R R^2 F**-value**
β **(Sig. T)**	0.256 (0.011)	-0.225 (0.023)	0.229 (0.022)	0.401 0.161 5.744

In words: The attractivity in this case will be formulated as a function of the parameters business areas (positive influence), greenbelts (negative influence) and the municipal areas (positive influence).

4. Purpose: home – education; mode: bike

No significant variables

5. Purpose: home – education; mode: bus

No significant variables

6. Purpose: home – education; mode: by foot

	population	**municipal**	R R^2 F **-value**
β **(Sig. T)**	0.339 (0.001)	0.274 (0.005)	0.417 0.174 9.575

In words: The attractivity in this case will be formulated as a function of the parameters population and the municipal areas (both with positive influence).

7. Purpose: shopping; mode: bike

	business	R R^2 F **-value**
β **(Sig. T)**	0.234 (0.023)	0.234 0.055 5.325

In words: The attractivity in this case will be formulated as a function of the parameter business areas (positive influence).

8. Purpose: shopping; mode: bus

	population	**Business**	**industry**	R R^2 F **-value**
β **(Sig. T)**	0.280 (0.006)	0.208 (0.043)	-0.204 (0.035)	0.511 0.261 10.578

In words: The attractivity in this case will be formulated as a function of the parameters population, the business areas (both with positive influence) and the industrial areas (negative influence).

9. Purpose: shopping; mode: by foot

	population	R R^2 F **-value**
β (Sig. T)	0.360 (0.000)	0.360 0.129 13.674

In words: The attractivity in this case will be formulated as a function of the parameter population (positive influence).

10. Purpose: return; mode: bike

No significant variables

11. Purpose: return; mode: bus

	population	**public**	R R^2 F **-value**
β (Sig. T)	0.602 (0.000)	0.280 (0.001)	0.637 0.406 31.043

In words: The attractivity in this case will be formulated as a function of the parameters population and the public areas (both with positive influence).

12. Purpose: return; mode: by foot

	population	R R^2 F**-value**
β (Sig. T)	0.542 (0.000)	0.542 0.294 38.245

In words: The attractivity in this case will be formulated as a function of the parameter population (positive influence).

9.2 Results in the Urban/Regional Sector

In the figures A3.2a, A3.2b, and A3.2c in appendix A3 the population distributions on the level of the 94 traffic cells (the so-called urban area) are depicted for the years 1987, 1996 and 2010. Apart from a total increase of the population numbers from year to year a re-distribution can be recognized in the early nineties, especially in the North and the South of the Urban Area. The population is concentrating in the centre of the Urban Area and shows a maximum of distribution in an East-West line through the city.

The results of the multiple regression are shown in table 9.3. The statistical tests show that the significance of the used variables is not very high.

On the base of the 94 traffic cells the following (independent) variables are used:

- average speed bike
- average speed bus
- average speed foot
- number of population reachable in 30 minutes by bike
- number of population reachable in 30 minutes by bus
- number of population reachable in 30 minutes by foot
- dwelling area
- business area
- public area
- industrial area
- greenbelts

- municipal area (gas works, water works, ...)
- storage area (storehouses)
- especially area (public leisure centers)
- external traffic area (railway stations, docks, etc.)

To consider the interactions between transport system and population distribution variables such as transport attractivities of the cells or travel times are used, too. The best result of the regression is the following:

Table 9.3: Best regression result, population analysis

	Transport Attractivities of the Purpose *shopping* weighted with the average speeds	**Transport Attractivi-ties of the mode *bike* and the purpose *home-work***	**External traffic**	**Business**	R R^2 F **-value**
β **(Sig. T)**	0.2877 (0.0082)	0.1755 (0.0966)	0.2373 (0.0226)	-0.1702 (0.1189)	0.4142 0.1715 3.00

10. Forecasting of the Traffic and Urban/Regional Development of Nanjing City for Different Scenarios

The present chapter describes three scenarios A, B and C. On the one hand we do not only want to simulate the traffic with respect to bikes, buses and pedestrians but also and above all the *car* traffic. On the other hand there are only rather limited data available concerning the cars. Under this data situation in subsection 10.1 a possibility of a simulation of the car traffic is worked out. This simulation is used in the basic scenario (scenario **E**) as well as in the three scenarios **A**, **B** and **C** which are described in detail in subsection 10.2. Subsection 10.3 gives an overview over the results of the simulations of scenarios A, B and C in comparison to the basic scenario E. The corresponding figures describing details of the scenarios are found in the appendix A4. In the text of the subsection 10.4 the references to the figures are given. Finally, in 10.5 the results of 10.4 are summarized and detailed conclusions are drawn which complement the more general conclusions and recommendations of chapter 11.

10.1 Simulation of the Car Traffic

The number of cars, buses and bikes can be estimated by a linear or non-linear forecasting. Here, a certain share of trips by bike and bus will be replaced by trips with a car. We continue assuming that the number of bikes and buses is constant (reference value of 1996) with a linear growth of private and firm cars on the base of the development between 1992 and 1996. We obtain for the modal split (see tables 10.1a and 10.1b as well as figures 10.1a and 10.1b)

- Modal split 1987: car – bike 0,8%
- Modal split 2010: car – bike 6,4%

Tables 10.1a and 10.1b: Number of different vehicles in Nanjing and the share in %

Category Year	Bikes	Buses	Private cars	Firm cars	Trucks
1987	1201299	1117	85	9335	21745
1988	1332761	1127	273	10546	23792
1989	1466991	1128	751	11317	24569
1990	1583099	1131	1388	11884	25469
1991	1171377	1170	1596	13255	26813
1992	1344956	1316	1915	13906	28837
1993	1523962	2412	2490	21529	33435
1994	1659769	2469	3112	26187	35236
1995	1846468	2321	3758	29998	38181
1996	1886667	2328	2498	35103	41961

Share in % Year	Bikes	Buses	Private cars	Firm cars	Trucks
1987	97,38	0,09	0,01	0,76	1,76
1988	97,39	0,09	0,02	0,77	1,74
1989	97,49	0,08	0,05	0,75	1,63
1990	97,54	0,07	0,09	0,73	1,57
1991	96,47	0,10	0,13	1,09	2,21
1992	96,69	0,09	0,14	1,00	2,07
1993	96,22	0,15	0,16	1,36	2,11
1994	96,12	0,14	0,18	1,52	2,04
1995	96,13	0,12	0,20	1,56	1,99
1996	95,73	0,12	0,24	1,78	2,13

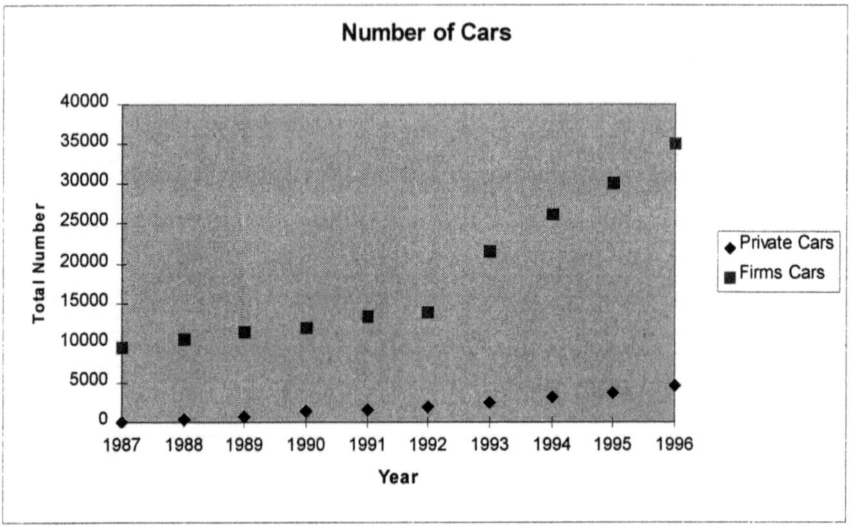

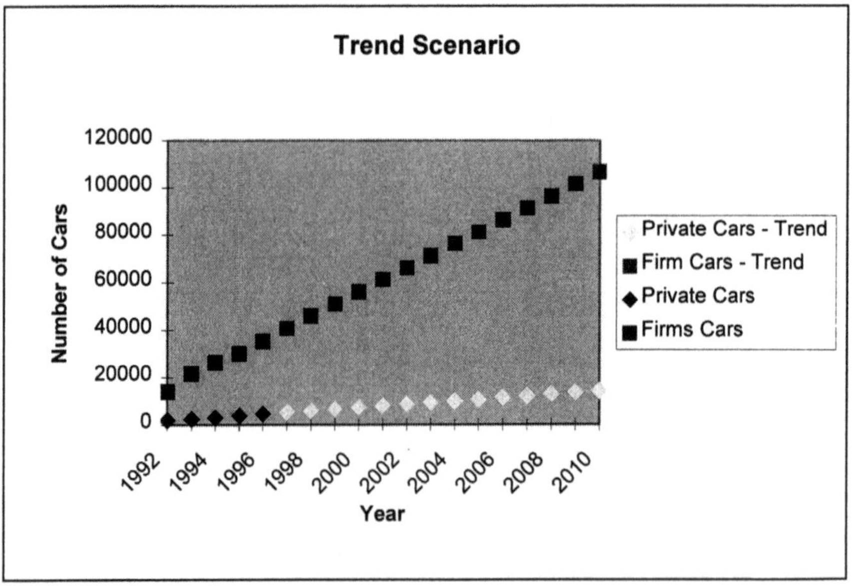

Fig. 10.1a and 10.1b: Estimated numbers of cars for the years 1987 - 2010

Since on the one hand we lack any actual data concerning the car traffic of Nanjing and on the other hand the 85 entries (this corresponds to a share of 0,01% of all vehicle types, see table 10.1a and 10.1b) of the OD-matrix *car* from the survey 1986 are clearly no sufficient basis to forecast the momentary car traffic situation in a reasonable way, we have to draw all conclusions out of the data concerning the trip modes bike and bus.

In order to obtain an OD-matrix for the trip mode car we make the following assumptions:

1. With a yearly increasing number of cars more and more road users switch from bike, resp. bus to the car. This switching process happens in dependence of the trip distance.
2. The sum of all entries of the resulting OD-matrix *car* for the first trip time interval (5h-10h) is approximately the number of cars being expected for the considered year (see tables 10.1a – 10.1b).

The implementation of these assumptions to our model looks as follows: We adapt the OD-matrix *bike* (*bus*, resp.) by subtracting $x_{ij}^{(bike)}(d_{ij})\,\%$ ($x_{ij}^{(bus)}(d_{ij})\,\%$, resp.) from each OD-matrix entry $F_{ij}^{(bike)}$ ($F_{ij}^{(bus)}$, resp.). The distance-dependent functions $x_{ij}^{(bike)}(d_{ij})$ and $x_{ij}^{(bus)}(d_{ij})$ are parables with the following structure:

$$x_{ij}^{(bike)}(d_{ij}) = x_{ij}^{(bike)}(0) + \frac{x_{ij}^{(bike)}(d_{\max}) - x_{ij}^{(bike)}(0)}{d_{\max}^2} \cdot d_{ij}^2$$

$$x_{ij}^{(bus)}(d_{ij}) = x_{ij}^{(bus)}(0) + \frac{x_{ij}^{(bus)}(d_{\max}) - x_{ij}^{(bus)}(0)}{d_{\max}^2} \cdot d_{ij}^2$$

where $x_{ij}^{(bike)}(0)$ ($x_{ij}^{(bus)}(0)$, resp.) is the percentage which switches from mode bike (bus) to car no matter how small the trip distance is, $d_{\max}$ is the largest distance between two traffic cells in the Urban Area and therefore $x_{ij}^{(bike)}(d_{\max})$ ($x_{ij}^{(bike)}(d_{\max})$) is the percentage switching to car when $d = d_{\max}$.

The sum of the subtracted parts of the OD-matrices *bike* and *bus* now build the newly generated OD-matrix *car*, so that the total number of traffic flows remains constant:

$$\sum_{i,j}\left(F_{ij}^{(bike,old)}+F_{ij}^{(bus,old)}\right)=\sum_{i,j}\left(F_{ij}^{(bike,new)}+F_{ij}^{(bike,new)}+F_{ij}^{(car)}\right)$$

The still undetermined parameters $x_{ij}^{(bike)}(0)$, $x_{ij}^{(bus)}(0)$, $x_{ij}^{(bike)}(d_{\max})$ and $x_{ij}^{(bus)}(d_{\max})$ can be calibrated in order to fulfil the second assumption:

$$\sum_{i,j}F_{ij}^{(car)}\approx 120.000$$

Here, we have a few degrees of freedom to fulfil this equation.
Note: Since in all the three following planning scenarios the values $x_{ij}^{(bike)}(0)$, $x_{ij}^{(bus)}(0)$, $x_{ij}^{(bike)}(d_{\max})$ and $x_{ij}^{(bus)}(d_{\max})$ are calibrated in such a way that the sum of trips by car is about 120.000, the average velocity as well as the total source/target traffic differs not too much in the three planning scenarios but one is able to see which relative consequences the one or the other decision would involve. The attempt to keep the values $x_{ij}^{(bike)}(0)$, $x_{ij}^{(bus)}(0)$, $x_{ij}^{(bike)}(d_{\max})$ and $x_{ij}^{(bus)}(d_{\max})$ constant has lead to a very high increase in car traffic in the scenarios B and C and could therefore not been compared with the basic scenario E and scenario A.

The above mentioned method of simulating the car traffic is used in all three planning scenarios A, B, C as well as in the basic scenario E (reference situation).

10.2 Description of the Scenario Technique and of the Scenarios E, A, B, and C

Overview over the Scenarios

The **scenario E,** *Empirically Given Basic Scenario,* describes the current situation of the transportation system and of the population distribution as developed in the previous sections with the distribution of areas according to the given data from 1990 (see figures A4.1a, A4.2a, A4.3a, and A4.4a in the appendix A4) and with the population of 1996 (see figure A3.2b in the appendix A3).

The **scenario A,** *the Mixed Land use Scenario,* describes the straight forward extrapolation of scenario E which amounts to a continuation of the given spatial mixture of housing and business land use. In other words, by this scenario one will be able to see the influence of the moderate increase of the total population and of the considerable increase of private car ownership within the 94 traffic cells. The utilization of areas in the Urban Area agrees with the utilization of 1990 as shown in the figures A4.1a, A4.2a, A4.3a, and A4.4a whereas the calculated population numbers for the year 2010 are depicted in fig. A3.2c.

The **scenario B**, *the Separated Land use Scenario*, describes a separation of housing and business areas according to plans for the year 2010 of the Nanjing municipal planning administration, depicted in the figures A4.1b, A4.2b, A4.3b, and A4.4b as well as a correspondingly calculated population distribution for 2010, (see figure A3.2c).

Scenario C, *the Separated Land use Scenario with Periphery Shifted Traffic*, describes, besides a periphery shifted traffic, e.g. engendered by an intense development of the industrial zones around the urban area of Nanjing.

In the latter scenario C all entries of the O-D-matrices *from* as well as *towards* the six main directions R1,...,R6 (see fig. A2.3 of Greater Nanjing) are increased in the following manner:
Analogously to the line of action in 10.1 (simulation of the car traffic), we make the changes distance-dependent. The number of trips F_{ij} depending on the distances d_{ij} for the trip modes bike and bus summed up over all trip purposes for the basic scenario E of 1987 show the following behaviour in average (but only for i or $j \geq 95$, corresponding to the directions R1,...,R6):
If the distance d_{ij} between one traffic cell and one direction or between two directions is less or equal 12.000 m, then

$$F_{ij} = 100 - \frac{100}{12.000m} \cdot d_{ij},$$

and we choose

$$F_{ij} = 0$$

for larger distances d_{ij}.

With this choice the right and the lower margin of the OD-matrices are filled up to simulate the through traffic in an appropriate way. The other entries of the OD-matrices are not changed and handled in the same way as in the planning scenario B, i.e. calculated with the planned utilization of areas as shown in the figures A4.1b, A4.2b, A4.3b, and A4.4b and with the population of 2010 as shown in figure A3.2c.

Evaluation - Procedure

Before describing the scenarios in detail we make a general remark about our procedure.

We will simulate the transport system with the modes car, bus, bike and foot, as well as the population distribution for different planning scenarios. To achieve this we have to take into account the interactions between both systems:

- The population distribution depends on accessibility measures, so that changes of travel times have effects on the population distribution.
- The traffic flows as well as the transport attractivities directly depend on the population numbers.

To forecast the coupled evolutions of the transport system and the population distribution for the different scenarios the following procedure will be applied:

- The share of car-owners is assumed to increase in the future. Therefore the modal split will change. We assume, that for short trips (up to a certain distance) X % of the trips by bike will be replaced by trips by car, for longer trips there will be a change from bus to car, so that Y % of the trips by bus will be made by car (for the calculation of X and Y see subsection 10.1). This procedure is done in all scenarios: the **basic scenario E** and the **scenarios A**, **B**, and **C**.
- The population numbers of the single traffic cells as well as the traffic behaviour of Nanjing depend on the distribution of building kinds such as houses, business, public, industry, greenbelt, municipal, storage over the cells. The figures A4.1, A4.2, A4.3, and A4.4 of the appendix A4 show the distribution of the four kinds industry, housing, business and greenbelt over the cells for the years 1990 and 2010. The discussion of the area-utilization is postponed to section 10.4 where its effect on the traffic evolution in the different scenarios is considered. To be able to make forecasts about the traffic evolution depending on the distribution of the areas two different cases are considered:

1. the existing area-utilization of 1990 with respect to the above mentioned classes is used in the **scenarios E** and **A**.
2. the planned area –utilization of 2010 with respect to the above mentioned classes is used in the **scenarios B** and **C**.

- A new bridge over the Yangtse in the North of Nanjing is planned as well as a new high speed train from Nanjing to Shanghai. In **scenario C** these aspects are taken into account by increasing the total through traffic in order to recognize potential changes in the population and in the traffic.

For these scenarios the population distribution as well as the travel flows for the different modes are calculated (see methodology in chapter 8). The travel flows have to be assigned to the network of the year 2000 to evaluate the effects of the different scenarios on the transport system.

The general framework for the scenarios in addition to the methodology so far (remember the figures 8.1, 8.2, and 8.3) is presented in figure 10.2.

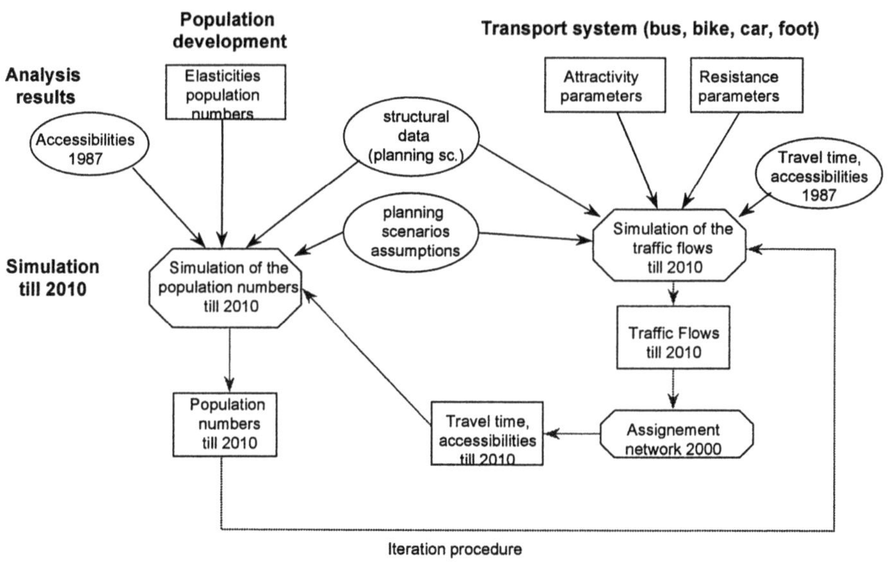

Figure 10.2: General framework for the scenarios

10.3 Analytical and Numerical Results of the Scenarios

All results are listed in the tables 10.2 and 10.3 as well as in the figures A4.5a – A4.5e for the basic scenario E, the figures A4.6a – A4.6e for the scenario A, the figures A4.7a – A4.7e for the scenario B, and finally the figures A4.8a – A4.8e for the scenario C. All these figures can be found in the appendix A4.

We shall present the following analytical and numerical results for all scenarios in order to get an overview over the characteristic behaviour of the calculated situation:

1. The number of iteration steps of each scenario which were necessary until the iteration procedure converged. These numbers are only of technical interest (see table 10.2).
2. The numbers $x_{ij}^{(bike)}(0)$, $x_{ij}^{(bus)}(0)$, $x_{ij}^{(bike)}(d_{\max})$, and $x_{ij}^{(bus)}(d_{\max})$ taken as described in 10.1. These values are measures for the distance-dependent numbers of trips in which the trip mode is changed from bike, resp. bus, to car (see also table 10.2).
   Remark: Since the scenarios A, B, and C are created in such a way that the total number of cars in the Urban Area is about 120.000 one has to count on *lower* values x_{ij} if more traffic is induced by the scenario.
3. The total capacity of the Urban Area traffic as a characteristic measure for the traffic development (see table 10.3).
4. The assignment of the Urban Area street network to detect the weak points of the capacities of the traffic mode car in more detail (see figures A4.5a, A4.6a, A4.7a, and A4.8a).
5. The source traffic of the 94 traffic cells plus the six main directions R1,...,R6 summed up each over all destination cells to see in which cells the most trips start (see figures A4.5b, A4.6b, A4.7b, and A4.8b). For the exact numbers of trips from the directions R1 – R6 see the table 10.4.
6. The target traffic of the 94 traffic cells plus the six main directions R1,...,R6 summed up each over all origin cells to see in which cells the most trips finish (see figures A4.5c, A4.6c, A4.7c, and A4.8c). For the exact number of trips in the directions to R1 – R6 see also the table 10.4.
7. The average speed (source) over all trips starting in each of the 94 cells to get a measure of attractivity of the cells with respect to the availability (see figures A4.5d, A4.6d, A4.7d, and A4.8d).

8. The population distribution in the Urban Area caused by the changes in order to compare the influences of the scenarios with respect to the migration behaviour of the inhabitants (see figures A4.5e, A4.6e, A4.7e, and A4.8e).

All figures in the appendix A4 show the Urban Area in the first trip time interval (5h – 10h). The colour illustrates the intensity of the examined aspect such as number of trips, number of population, etc.. In the figures A4.5c – A4.8c and A4.5d – A4.8d showing the source, resp. target, traffic green arrows are added which point to the directions R1 – R6. The width of these arrows are a measure of the trip numbers.

Table 10.2: Number of iteration steps and characteristic values for changing from bike (bus) to car for the scenarios

Scenario	Number of iterations	$x_{ij}^{(bike)}(0)$	$x_{ij}^{(bus)}(0)$	$x_{ij}^{(bike)}(d_{\max})$	$x_{ij}^{(bus)}(d_{\max})$
Basic scenario E	1	15%	15%	5%	5%
Scenario A	1	15%	15%	5%	5%
Scenario B	2	12%	12%	4%	4%
Scenario C	2	10%	10%	4%	4%

Table 10.3: Total traffic capacity and the relative increase compared with the basic scenario

	Total traffic capacity per person (km/5h)				**Relative increase to the basic scenario**		
Trip mode	**Basic sc.**	**Sc. A**	**Sc. B**	**Sc. C**	**Sc. A**	**Sc. B**	**Sc. C**
Car	349.924	471.775	513.888	513.094	34,8%	46,9%	46,6%
Bike	1.154.680	1.582.185	5.857.322	6.173.513	37,0%	407,3%	434,7%
Bus	1.027.899	1.364.516	1.460.230	1.690.888	32,7%	42,1%	64,5%

Table 10.4: Number of trips from and towards the directions R1,...,R6 during the first trip time interval 5h – 10h

	Basic scenario E		Scenario A		Scenario B		Scenario C	
Direction	From	Towards	From	Towards	From	Towards	From	Towards
R1	149	123	228	93	241	114	1848	1799
R2	148	87	271	132	287	66	1714	1528
R3	246	101	149	122	153	123	2594	2431
R4	262	339	304	306	323	335	1853	1902
R5	145	57	275	119	290	121	2080	1937
R6	442	96	228	146	239	150	1982	2017

10.4 Discussion of the Scenarios

The discussion of the scenarios plays a central role in the present study. On the basis of the figures shown in the previous subsection as well as of the calculated traffic capacities one is now able to point out which changes would be caused (on the level of the transportation as well as on the urban/regional sector) by the planned measures.
Only looking at the adaptation of the population numbers to the year 2010, i.e. without any change of the utilization areas (scenario A), the general situation only changes a little, whereas the situation is totally different with respect to the scenarios B and C:
Here, the total driving capacity raises by more than 400% (see table 10.3). In other words the planned utilization changes would induce very heavy traffic. In detail:

The planned changes of utilization

The figures A4.1a and b, A4.2a and b, A4.3a and b, and A4.4a and b clearly show that Nanjing is planning drastic changes between the years 1990 and 2010, especially within the CBD (central business district) which is planned to become a business district in its true sense: no industry, almost no lodgings.

Industry (fig. A4.1a, A4.1b)

It is planned to transfer the complete industry of Nanjing from the CBD to the large industrial zones A_1, A_2, B, and C and to the outer traffic cells of the Urban Area. The large industrial zones partially do already exist, they are under construction. The industrial sector within the Urban Area is planned to decrease from 20,3 km^2 (1990) to 17,7 km^2 which means a decrease of 12,8%.

Housing (fig. A4.2a, A4.2b)

The lodging area is subject to the same planning-principle as the industrial area: To keep the CBD free of flats as far as possible. This can be seen in the planned transfer of the main housing areas to the North-east as well as to the South-east of the Urban Area, whereas currently a 'housing belt' extends from the East through the CBD. The housing sector within the Urban Area will increase from 33,9 km^2 to 65,0 km^2 (=91,7%) according to the plan. Since the main industrial zones A_1, A_2, B, and C lie outside of the Urban Area where only less people are living, the plam follows the principle of separation of landuse between industry and housing.

Business (fig. A4.3a, A4.3b)

The planned changes in the business sector go in the same direction. With a total planned increase of more than 200% of the business area (from 3,6 km^2 to 10,9 km^2) one has to be prepared for an immense increase. The bulk of this increase will be located in the CBD. We remark that the CBD is planned to contain one sector only, namely the business sector, whereas bureaus, offices, etc. are planned to be transferred to the outer cells of the Urban Area. Once more the planned *separation* of land use can be clearly seen.

Greenbelt (fig. A4.4a, A4.4b)

The planned adding of greenery to the city area in Nanjing is a further interesting aspect. The figures 10.6 give account of it. An increase of the greenbelt of 41,4% (from 16,9 km^2 to 23,9 km^2) distributed over the whole Urban Area is planned.

It should be mentioned that all depicted areas are ground areas and not floor space areas.

The scenarios

Equipped with the knowledge of the planned utilization we are now able to compare the three scenarios A, B, C with the basic scenario E (fig. A4.5a – A4.5e):

As expected, the **scenario A** (fig. A4.6a – A4.6e) leads only to some quantitative but not to deep qualitative changes as compared to the basic scenario E concerning nearly all aspects. These changes are of almost the same or even lower order than the relative changes in the total population numbers. The total population number increased from 1990 (1.508.917 persons in the Urban Area) to 2010 (2.284.303 persons) by more than 50%. The capacity figures A4.5a and A4.6a clearly demonstrate the weak points of the traffic network which already existed before. This refers firstly to the CBD and secondly to the extension to the North-west towards the bridge over the Yangtse river: In the scenario A we find a continuous connection of overloaded streets between the South-east of the CBD and this bridge (with more than 5.000 cars per 300 min. in the first trip time interval).
The average speed in the scenario A is with 17,0 km/h slightly slower than in the basic scenario E with 17,3 km/h. The average speed (source) is calculated in the following way:

$$\bar{v} = \frac{1}{100^2} \cdot \sum_{i,j} \frac{d_{ij}}{t_{ij}} .$$

In addition, there is a remarkable analogy between the source traffic in the scenario A (fig. A4.6c) and the calculated population numbers of 2010 (fig. A3.2c) as well as between the corresponding figures for the basic scenario E (fig. A4.5c) and the population numbers of 1986 (fig. A3.2a). On the one hand, this of course is an indicator of the consistency of the ITEM model, but on the other side these effects can be used to make first approximations.

Since in the basic scenario E as well as in the scenario A the principle of mixed land use is still fulfilled and the large industrial zones in the outskirts have not yet been erected, the flows to and from the Urban Area (the directions R1,...,R6) remain low.

Finally, the population distribution (fig. A4.6e) induced by the scenario A seems to be determined by the expected population of 2010 and therefore yields an expected result.

The first remarkable point in the **scenario B** (fig. A4.7a – A4.7e) is the very low total average speed $\overline{v}$ of 16,4 km/h. In more than the half of all traffic cells in the Urban Area the average speed (source) lies under 16.6 km/h. Remarkable is also the much higher total capacity of the streets (see table 10.3). In more than the half of all traffic cells in the Urban Area the average speed (source) lies under 16.6 km/h. Of course these cells lie in the CBD and around it. In contrast to the basic scenario E and to the scenario A the regrouping in the CBD induces heavy source traffic, especially in the North and West of Nanjing: Whereas in scenario A in the total East of the Urban Area less than 200 trips per cell were calculated for the 5 hours (5h – 10h), the utilization plans generate cells with 2.000 trips and more; with respect to source traffic as well as to target traffic. What was to be expected has been confirmed: That there exists a clear correspondence between

Planned area (housing)	↔	Source traffic
Planned area (industry/business)	↔	Target traffic

These relations are evident, in particular in view of the considered time interval (5h – 10h) when most people are driving from their home to their working place. Since in the capacity figure A4.7a another scale is used than in the other capacity figures because of the very large traffic volume from and to the cell 93 and 92 in the North, it is better to direct one's attention to the colours in which the streets are depicted. Then we recognize an extension of the critical traffic zones of the Urban Area to the North-east. This corresponds to the planned areas for the lodgings (fig. A4.2b).

The increase of traffic over the Yangtse river from the outskirts towards the Urban Area finds its explanation in the large industrial zones in the North-west of the city. The large industrial zones in the outskirts are not taken into account in scenario B, but indirectly in scenario C where the reasons for the increase of through going traffic can be explained by the industrial zones. In the present situation the most trips have the direction R4 (South-eastern direction); this direction leads to Shanghai.

The population re-distribution in the scenario B can be understood as the consequence of the new distribution of the population in 2010 and the new distribution of housing area in 2010. That means in the CBD live less people than in the scenario A and in the basic scenario E.

The results of **scenario C** (fig. A4.8a – A4.8e) are more or less comparable with the results of scenario B. The main difference is that the shifted traffic, after standard assignment, leads to congestions on specific primary roads, in particular on the main outlets (in the directions R1,...,R6). Here, an increase of more than 400% can be recognized.

Summarizing it can be said that the results of the scenarios B and C clearly show which traffic problems Nanjing would face if it would fully realize the utilization of the areas as shown in the figures (A4.1 – A4.4), according to the principle of the separation of landuse. Since on the other hand this plan of erecting large industrial zones in the outskirts instead of the Urban Area of Nanjing intends to enhance the quality of life and since these zones are already under construction the only way out of the dilemma seems to be to build a second bridge over the Yangtse river in the North of Nanjing.

11. Summary and Recommendations

By the year 2000, more than half of the world's population will live in cities. This means that not only an *extensive exchange* of population, goods and information in a globalised world within this system of cities can be expected but also that the transport of goods, population and information must be *effectively managed.* A city which wants to represent an important node in this network must provide beside appropriate economic, social and cultural conditions and political stability, qualified labour, and an urban as well as an internationally operating interurban system of transport. Moreover, all cities face a common problem: they must possess the capacity to sustain unprecedented numbers of citizens within limited budgets and severe environmental constraints.

One of those emergent nations facing a rapid urbanisation process is **China**. China faces a substantial increase in population, to an estimated 1.3 billion by the year 2000, and to 1.5-1.6 billion by the year 2050 (State Planning Commission, 1994). For all Chinese commentators these statistics dominate every consideration of global development.

Nanjing, the former capital of Southern China is the *regional capital* of the Jiangsu Province. It has 2.61 million inhabitants and is therefore one of the big cities in China. The city of Nanjing (the old capital of six dynasties of ancient southern China) is located about 300 km to the north-west of Shanghai on the Yangtse-River (31° north latitude, 118° east meridian). Nanjing City is an important traffic node, since it has the biggest river port in China that links Nanjing with the ocean. Furthermore, in Nanjing there is one of the few bridges crossing the Yangtse River. Hence, it connects Xuzhou (and farther away Beijing) in the north and Zhengzhou in the north-west, Hefei (the capital of Anhui Province) in the west, and Wuxi, Suzhou and Shanghai in the east. Its spatial location underlines its strategic position within the global transport network of China.

In **Greater Nanjing** nine industrial development zones exist. These nine development zones are located outside the urban area. Hence, they have to be accounted for by the six mail traffic directions. For the traffic situation in the urban area those industrial development zones are most important that are located nearby the exit points of the various directions.
According to these criteria, the following **industrial zones** have been considered in detail (see Figure A2.2, appendix A2):

1. The Headquarter of the Nanjing High and New Technology Industry Development Zone.
2. The Nanjing Economic-Technical Development Zone including the Nanjing Xingang High and New Technology Industry Park.
3. The Nanjing Jiangning Economic and Technological Development Zone including the Nanjing Jiangning High and New Technology Industry Park and the Nanjing Non-State-Run Science and Technology District.

The volume of *passenger transport* generated by the location of work evidently heavily depends on where the employees live. Furthermore, the amount of freight traffic generated depends on the transport intensity of the production process and on the composition of the firms located in an industrial development zone.

The above mentioned problems are the motivation for the **ITEM project**. ITEM (Integrated Transport and Evolution Model) analyses the traffic as well as urban development in emergent nations based on integrated mathematical tools. The interaction between urban and regional development and transport infrastructure is the main target of the analysis. In the light of the many uncertainties concerning future transport systems as well as population development, it has proved to be valuable to employ scenarios as a vehicle for the communication with policy-makers. Scenarios are richer in scope than conventional forecasting tools and offer more innovative investigation to policy-makers.

A **scenario** is defined as a tool describing pictures of the future world within a specific framework under specified assumptions. The scenario approach employed includes the description of four scenarios designed to compare and examine alternative futures with respect to traffic development and population redistribution.

The usefulness of scenarios in a *planning context* consists in the attempt to shape the future taking into account different scenario results, instead of just adapting to what may emerge. This means, that scenarios can provide recommendations and guidelines in the political decision process. Such methodologies are even reasonable when essential parts of the system under study cannot be controlled by policy measures and depend on exogenous factors.

External factors influencing the transport sector of Nanjing are e.g. global institutional/political developments inside (or outside) China, or the introduction of new technologies.

In the post-war period we observe a steady increase of mobility, not only for daily home-to-work trips, but also for business trips or leisure trips. In the age of globalisation it seems likely that a continued pressure will exist towards higher mobility levels.

In this context, it should be emphasised that mobility as such is not bad, because it creates many economic benefits to both consumers and producers. The problem is however, that there are many social costs involved in the form of pollution, accidents, traffic congestion etc. which are not (fully) charged to the sources of the externalities. Thus, the main question for many governments nowadays is whether it will be possible to ensure that mobility rates are compatible with sustainability criteria. This would mean that the so-called de-coupling hypothesis would have to be implemented, which means a de-coupling of economic growth and environmental charges caused by the transport sector. This would of course require drastic changes in our modes of production, consumption, transportation and technological innovation. Clearly, there are many uncertainties involved in developing such new (policy) strategies, and hence it seems wise to rely on appropriate scenarios which might depict some of the future developments and allowing us to identify possible bottlenecks.

Of course, a possible de-coupling of economic growth and transport infrastructure requires at least a minimal transportation system and an effective information network. Only under those conditions a city or region can maintain its importance in the global economy.

In the Chinese context, it is clear that decisive *external factors* for scenarios depend on the demography, the economic policy (e.g., the open door policy), as well as on the settlement and industrialisation policy. Thus, these factors of uncertainty will influence the transportation system in China, on the urban level as well as on the national level.

At this point we summarize the main results of scenarios E, A, B, C:

Scenario E, *the Empirically Given Basic Scenario*, describes the current situation of the transport system and of the population distribution purely based on empirical data.

In the subsequent scenarios A, B and C the same considerable increase from 1990 until 2010 of private car ownership and the same moderate increase of the total population numbers have been assumed.

Scenario A, *the Mixed Land use Scenario*, describes the straight forward extrapolation of scenario E which amounts to a continuation of the given spatial mixture of housing and business land use. The result is that no dramatic changes occur concerning the transportation system. In this respect the mixed land use proves to be advantageous .

Scenario B, *the Separated Land use Scenario*, describes a separation of housing and business areas according to plans for the year 2010 of the Nanjing municipal planning administration. The result is that the traffic load in the centre and North of Nanjing increases dramatically. This means: Increase of private car ownership in combination with separation of landuse leads to a partial overcharge of the transportation network.

Scenario C, *the Separated Landuse Scenario with Periphery Shifted Traffic*, describes, a periphery shifted traffic, e.g. engendered by an intense development of the industrial zones around the urban area of Nanjing. The result is that the shifted traffic, after standard assignment, leads to congestions on specific primary roads, in particular on the main outlets.

Recommendations

A comparison of the different scenarios clearly demonstrates that the development of a strong CBD with many jobs located in the centre of Nanjing and with a transfer of housing units to the periphery of the City generates in the long run a tremendous amount of traffic, mainly in the morning and evening hours. Therefore planners should avoid a strict separation of the spatial location of workplaces and housing/living opportunities. On the other hand synergy aspects have to be considered as well. This means a concentration of services and light industries in the CBD bears many synergies because of a high accessibility of different agents/firms. This high accessibility levels are related to the short distances.

What should the planners do in this conflicting situation?

- Nanjing should develop towards a multi-centre municipality instead of a single-centred central place.
- Each of the different small centres of the metropolitan area should provide all facilities needed for daily living: Local administrative offices, small shops, housing units, market places, cinemas, theatres and so on. This means within those small centres a good mixture of different services should be provided.
- The accessibility of the different centres of the metropolitan area of Nanjing must be considerably improved. Therefore, the capacities of the corresponding traffic network to the CBD must be increased in order to be able to manage the growing transport demand, especially during the rush hours.
- The expansion of the traffic network should have the same priority as the development of the CBD in order to avoid intermediate phases of chaotic and costly traffic jams.
- The expected increase of the volume of transit traffic from the south of China (Hongkong) to north (Beijing) and from the east (Shanghai) to west (Hefei) requires the development of a very effective ring road around the City of Nanjing. In order to avoid long time traffic jams on this ring road we recommend the reduction of the number of possible exits (junctions) considerably. Only with very few and well designed exits it is possible to keep a main part of the *intra urban traffic* outside of the ring road.
- As our simulations demonstrate, the second bridge over the river Yangtse which is already under construction is certainly an important step toward an effective management of the transit traffic and of the inter urban traffic.
- The urban traffic crossing the river Yangtse should be minimized. This should be an objective of urban/regional policy. An adequate mixture of workplaces, services and housing units at both river banks is therefore recommended.
- The transport system of Nanjing bears some bottlenecks with respect to the transport modes. Especially the public transport system (mainly busses) must be considerably improved. The development of the public transport system of Curitiba, the capital of the state of Paraná in the south of Brazil, could be partially used as a guideline (cf. Rabinovich and Leitman 1996). The essentials of the urban traffic system of Curitiba are:

- Busses cover the main part of the public transport because subways are too expensive
- Separate bus lanes on the main roads
- Efficient bus entering and bus leaving system realized by bus stop stations with automatic ticket sales
- The bus companies are private and paid for the covered distance per day and not for the number of sold tickets

- A consequent separation of the modes by foot, bike, busses and cars for the main roads should be organised. The typical existing main roads of Nanjing have already a structure which could easily be transformed into a more effective road system. In any case it is not necessary to cut down the beautiful old trees which separate already different lanes (tracks) and which represent one of the nicest and best known sights of Nanjing.
- The management of car parking within Nanjing bears some further innovative intervention possibilities. Of course there are multi-storey car park possibilities needed in the CBD with high accessibility. On the other side Park-and-Ride opportunities on the periphery of the City could help to reduce the traffic volume in the centre. However, Park-and-Ride works effectively only, if the transaction costs from private car to public transport system are less than the private travelling costs. The choice of an appropriate tariff structure is needed. New telematic tools should be used to give at least estimates of expected waiting time and number of vacant parking places. The public authorities should develop a uniform Park-and-Ride management structure for Nanjing in order to support its acceptance. Moreover, intermodality aspects should not only consider the mode-change from car to bus but also from car to bike.

The above mentioned development aspects of the urban and the transport system of Nanjing have to be seen in the light of the rapid changes in the field of transport and communications techniques. Several trends combine to diminish the impacts of transport infrastructure on regional development and are therefore helpful in decoupling economic growth and transport infrastructure:

- Telecommunication may reduce the need for some goods transport and person trips. However, they may also increase transport by their ability to create new markets.
- With the shift from heavy-industry manufacturing to high-tech industries and services other location factors have become relevant and replace partially the traditional ones. These new location factors include factors concerning leisure, culture, image and environment, i.e. quality of life, and factors concerning access to information, high-level services, qualified labour and to municipal institutions.

The city of Nanjing is in a fortunate situation in so far as it is in a transitional and fast developing phase. This situation allows of realizing many innovative ideas concerning an integrated city and traffic development. The recommendations drawn from the results of the ITEM project indicate that good chances exist for Nanjing to arrive at a sustainable city evolution guaranteeing a sufficiently high level of mobility.

11.1 Achnowledgement

We thank the DaimlerChrysler AG for sponsoring our project by a generous financial grant and by providing all means for implementing our scientific project in full independence.

We gratefully acknowledge the intense co-operation with our partners Prof. Wang Wei and Prof. Deng Wei from the Southeast University of Nanjing.

Finally, we are grateful to our Senior-Advisors Prof. Dr. Dr. h.c. mult. Ilya Progogine and Prof. Dr. Dr. h.c. mult. Hermann Haken for their continuous interest and supervision of the project.

Appendices

Detailed Mathematical Description of the Model, Figures of the Results

Appendix A1 Details of the Integrated Transport and Evolution Model

Traffic demand is the result of social and individual activities as well as the evolution of land use patterns (SACTRA 1994a). Accordingly, individual decision processes of the agents form the basis (individuals, companies etc.) of all transport phenomena. In this way, the traffic volume is influenced by the total number of activities, by the places and starting time of activities, by the means of transport and route selection, by the co-ordination of activities with other agents (e.g. formation of car pools) as well as by feedback effects.

In total therefore, traffic demand is the result of a great number of individual decision processes. These complex processes can take place on a short-term or long-term basis. In order to represent the demand behaviour of the agents in a model one must take into account the following:

> *„The important point here is that travel demands on any particular route or network are not pre-ordained. They arise out of tens of thousands of individual choices being made every day, with each person in the population deciding where, when and how they wish to travel. Quite reasonably, they make these decisions in their own interest, rather than that of the community as a whole, and on the basis of the best information available at the time, which is often rather poor. The purpose of building a mathematical model of the travel demand that results from all these choices is to be able to predict how that demand might change, as the circumstances determining those choices change." (SACTRA 1994a, p.15)*

In traditional traffic models changes in the place of work or the locations of firms and infrastructure can only be considered by exogenous variables (GAUDRY/ MANDEL/ ROTHENGATTER 1994). In the ITEM traffic model the interdependence between the urban/regional subsystem and the traffic subsystem is considered. This traffic model comprises also induced traffic components and can only so far be compared with usual gravitational approaches, as the traffic flow among other things depends on the respective traffic resistance between source and target areas.

It is our aim to derive the trip frequency via a behaviour oriented decision model. Behaviour of individuals and households is described on the basis of sub-populations. These groups are characterised by comparable socio-economic features so that the traffic behaviour of the individual decision makers within the same group is approximately uniform, however it is quite different between individuals of different groups. Correspondingly, the entire population of the region under investigation is subdivided into groups (KUTTER 1972, MEIER 1990). The activity patterns and the corresponding trip distribution can therefore be taken as group-related.

Depending on time of day and calendar day, the traffic behaviour of the individual agents is determined by different activity patterns. With respect to mobility the traffic behaviour of the individuals in general represents merely an advance payment so that other activities of the person can be facilitated. Therefore, the trip frequency is the result of a great number of "individual" decision processes and therefore related to rational and irrational motives of the individual agents. It is assumed that for each trip to a traffic cell purpose-specific attractivities can be assigned depending on specific characteristics of the traffic cell (number of apartments, work places, shopping possibilities, leisure time facilities and so forth). Both the respective cell specific socio-economic factors as well as traffic demand and supply dependent factors are important for these decision processes. For this reason a high degree of „feedback" between the traffic system and urban/regional system can be expected (NIJKAMP, REGGIANI, 1992).

Furthermore, since these different activity patterns of the individuals or households vary strongly over the day, the corresponding regional attractivities show a strong time dependence as well. Therefore, the transportation system is not in equilibrium with respect to time of day.

The decisions of the individual agents are in general not independent of each other. Rather, the spatial distribution of these agents play a decisive role in addition to other socio-economic factors. Thus pigeon effect, band-wagon effect, synergy effect or group effect results in a dependence of the attractivities of the traffic cells on the distribution of the agents and has to be considered.

In this way, time-dependent attractivity differences between the traffic cells and distance resistances are considered to be essential performance-influencing factors of the traffic flows. Moreover the uncertainty in the decision process due to e.g. insufficient information about the characteristics of the respective traffic cells has to be included. For instance, a household is better informed about shopping possibilities in the local environment than about those possibilities in traffic cells far away, which are therefore not considered in its decisions. This is considered by cell specific accessibilities, which can contribute to the attractivity level of a particular cell.

A1.1 The Micro Level

The micro level of the society is determined by the individual decision behaviour of the different agents of the economic system. So-called "attitudes" are in the foreground of every individual concerning specific "aspects" of the society, such as the present earnings, the kind of work, the present living conditions, the family background, leisure facilities, politics, the traffic situation, to mention a few performance-influencing factors. Let

$$p_{ij}^{\alpha r(l)}(\vec{E},\vec{M},\vec{\kappa}_l) \tag{A1.1}$$

be the probability per unit of time that an individual l, ($l = 1,.., I$) carry out a trip from area i to area j, given a certain population distribution $\vec{E}$ and distribution of dwellings, companies, shopping centres $\vec{M}$ in the considered region. The individuals might be allocated to trip purpose specific subpopulations α. The different attitudes of each individual are summarised in the „attitude vector" $\vec{\kappa}_l$. The corresponding individual transition rates $p_{ij}^{\alpha r(l)}(\vec{E},\vec{M},\vec{\kappa}_l)$ can be determined e.g. via panel data (Courgeau 1985).

The Fishbein/Ajzen model is based on the hypothesis that a relation exists between the actual behaviour of an agent (e.g. during a change of the place of home) and his attitudes (FISHBEIN/AJZEN, 1975). However, empirical behaviour research shows that attitudes and actual behaviour must by no means be correlated (LAPIERE, 1934). A generalization of the Fishbein/Ajzen model includes habit effects and alternative possibilities as further determining factors for the actual behaviour. However an answer to the basic question

„Under what conditions do what kinds of individuals show what kind of behaviour?"

is still by no means possible. (FAZIO/ZANNA, 1981).

Dynamic micro models, based on panel data, have the advantage of being able to represent the modification of policy relevant variables in the course of time. However, modelling of traffic flows requires the preparation of a comprehensive (representative) data base, combined with corresponding costs.

A1.2 The Macro Level

The number of the system variables necessary for the description of the society is reduced drastically if one passes to the macro level.

The region under investigation is subdivided into L non-overlapping traffic cells i, $i = 1,...,L$. The total population might be further disaggregated (persons, budgets and so forth) into trip purpose related subpopulations α, $\alpha = 1,.., P$. The following assignment refers to the available data base

Trip purposes:
- $\alpha = (1)$ home - work
- $\alpha = (2)$ home - education
- $\alpha = (3)$ home – shopping
- $\alpha = (4)$ return

and so on. The introduction of further trip purposes depends on the data base of the region under consideration.

A subdivision according to traffic modes r, $r = 1,.., R$, is also introduced, since modifications of the modal split have to be considered and evaluated as well.

Traffic modes:
- $r = (1)$ on foot
- $r = (2)$ bicycle
- $r = (3)$ bus

Further traffic modes are also conceivable and possible.

The number of trips for the trip purpose α from traffic cell i to traffic cell j at time t (hour of the day) using traffic mode r is denoted by $F_{ij}^{\alpha r}(t)$. Then

$$F_{ij}^{\alpha}(t)=\sum_{r=1}^{R}F_{ij}^{\alpha r}(t) \tag{A1.2}$$

is the number of trips between i and j for trip purpose α and

$$F_{ij}^{r}(t)=\sum_{\alpha=1}^{P}F_{ij}^{\alpha r}(t) \tag{A1.3}$$

is the number of trips between i and j using mode r, and

$$F_{ij}(t)=\sum_{r=1}^{R}\sum_{\alpha=1}^{P}F_{ij}^{\alpha r}(t) \tag{A1.4}$$

is the total number of trips between i and j. Therefore, accordingly the total number of trips in the system is given by

$$F(t)=\sum_{i=1}^{L}\sum_{j=1}^{L}F_{ij}(t)\cdot \tag{A1.5}$$

The traffic volume $O_i^{\alpha}(t)$ (source volume) of the traffic cell i is gained by summation from the traffic flows:

$$O_i^{\alpha}(t)=\sum_{j=1}^{L}F_{ij}^{\alpha}(t) \tag{A1.6}$$

The number of trips $D_j^{\alpha}(t)$ for the trip purpose α into the traffic cell j at time t reads correspondingly:

$$D_j^{\alpha}(t)=\sum_{i=1}^{L}F_{ij}^{\alpha}(t)\quad\cdot \tag{A1.7}$$

The population distribution is denoted by

$$\vec{E}=\{E_1,\dots,E_i,\dots,E_L\}, \tag{A1.8}$$

where E_i is the number of agents of traffic cell i. E_i will be modified by the decisions of the individual agents to carry out a trip between the cell i and anyone of the other cells. Therefore, the population distribution $\vec{E}$ is connected via traffic related activities of the agents with a great number of individual decision processes and finally with the attractivities of the specific traffic cells.

The distribution of apartments, shops, companies, service facilities in the regional system are described via the stock vector

$$\vec{M} = \left\{ M_1^1, \ldots, M_L^1, \ldots, M_i^h, \ldots, M_L^{\tilde{H}} \right\} \tag{A1.9}$$

where the index h, $h = 0,.., \tilde{H}$ distinguishes different types of buildings according to the subsequent categories:

Building types:

- $h = 0,.., h1$ housing units of different kind (e.g. subdivided after size, equipment)
- $h = h1{+}1,.., h2$ firms of different kind (e.g. offices, enterprises)
- $h = h2{+}1,.., h3$ leisure time facilities of different kind
- $h = h3{+}1,.., H$ consumer markets, shopping centres, retail businesses
- $h = H{+}1,..,$ school types

In the individual traffic cells the stock values of the different building types change due to rational and emotional decisions, aspects and considerations of households and investors. In this case, a great number of socio-economic factors hardly to be estimated in their respective effects and in particular in its mix may play an important role. An essential task of the model evolution consists in the selection of those factor combinations which determine essentially the dynamics of the traffic and urban/regional level. In this case, the integrated transport and evolution model (ITEM) should be built up as simply as possible and it should be adapted to the available data base. According to this strategy the model is to be developed. Of course, there will always be points of criticism in the sense of disregarding certain effects which might be of importance in specific environment. Those effects which seem to be essential for the dynamics of the integrated model and which are not yet contained can be included via possible extensions and further refinements or disaggregations of the model. However, the availability of a corresponding data base is the essential point in the determination of an appropriate disaggregation level .

A1.3 Interactions between the Micro Level and the Macro Level

The micro level (individuals, households, companies) appears prima facie to determine the dynamics of the macro level completely, while there is no feedback of the macro level to the micro level. However, this is not the case. Rather, a mutual influence of the two levels occurs: Among other things, the actions (activities) of the individual agents of the economic system express themselves in the dynamics of the traffic flows and therefore in the time-dependent population redistribution of the system. Therefore, micro behaviour and macro dynamics in themselves are strongly coupled. Mathematically the dependence of the individual trip decisions on the macro state appears in the fact that the attractivities of the traffic cells are among other things dependent upon the respective macro state, characterised e.g. by the population distribution.

A1.4 The Stochastic Framework of the ITEM Model

In addition to rational motives, uncertainties have also to be considered in the case of decision processes. Therefore, a stochastic consideration of these processes is reasonable. For this reason the configurational probability $P(\vec{E},\vec{M},t)$ to find a certain regional distribution of population $\vec{E}$, households and firms $\vec{M}$ at time t provided the diverse reciprocal interactions of the actors of the economic system among one another. Therefore, the individual actors are not considered to be independent of each other.

For formal reasons, it is reasonable to introduce the socio-configuration $\vec{N}$,

$$\vec{N}=\left\{\vec{E},\vec{M}\right\} \tag{A1.10}$$

comprising the distribution of population and stocks at a given time t. The individual decision processes of agents will now have to be related to the temporal variations of the socio-configuration $\vec{N}$. Then

$$P(\vec{N},t)=P(\vec{E},\vec{M},t)\,. \tag{A1.11}$$

Of course, $P(\vec{N},t)$ must satisfy at all times the probability normalization condition

$$\sum_{\vec{N}} P(\vec{N},t)=1 \tag{A1.12}$$

where the sum extends over all possible socio-configurations $\vec{N}$. If the configurational transition rates $F_t(\vec{N}+\vec{k},\vec{N})$ from any $\vec{N}$ to all neighbouring $\vec{N}+\vec{k}$ are given, an equation of motion for $P(\vec{N},t)$ can be derived (e.g. see WEIDLICH/HAAG, 1993). It reads:

$$\frac{d}{dt}P(\vec{N},t)=\sum_{\vec{k}} F_t(\vec{N},\vec{N}+\vec{k})P(\vec{N}+\vec{k},t)-\sum_{\vec{k}} F_t(\vec{N}+\vec{k},\vec{N})P(\vec{N},t) \tag{A1.13}$$

where the sum on the right hand side extends over all $\vec{k}$ with non vanishing configurational transition rates $F_t(\vec{N}+\vec{k},\vec{N})$ and $F_t(\vec{N},\vec{N}+\vec{k})$, respectively. The eq. (A1.13) is denoted as a master equation and has a very direct and intuitively appealing interpretation. The change in time of the probability of the socio-configuration $\frac{dP(\vec{N},t)}{dt}$ is due to two effects of opposite direction, firstly to the probability flux from all neighbouring configurations $\vec{N}+\vec{k}$ into the considered configuration $\vec{N}$ namely $\sum_{\vec{k}} F_t(\vec{N},\vec{N}+\vec{k})P(\vec{N}+\vec{k},t)$ and secondly to the probability flux out of the configuration $\vec{N}$ into all neighbouring configurations $\vec{N}+\vec{k}$, namely $\sum_{\vec{k}} F_t(\vec{N}+\vec{k},\vec{N})P(\vec{N},t)$. Consequently the master equation represents a balance equation for probability fluxes. The transition rates occurring in the master equation (transition probabilities per unit of time) are associated directly with the short-term evolution of the conditional probability. For further explanations see (WEIDLICH and HAAG, 1983). The solution of the master equation, namely the time-dependent distribution contains in detailed manner all information about the population distribution process, about the evolution of the urban/regional level and therefore indirectly also about the trip frequencies. The configuration assigned to the time-dependent maximum of the distribution represents the most probable population distribution and/or the most probable distribution of the population and of the stocks in the urban/regional system at the time t, for given economic constraints.

The probability distribution $P(\vec{N},t)$ contains such a tremendous amount of information compared with the empirical information (data base) available concerning transport events and/or urban/regional processes that a less exhaustive description in terms of mean values seems to be adequate in most applications. Therefore, it is highly desirable to derive self-contained equations of motion for the mean decision behaviour of agents under given boundary conditions of the system. The mean population in traffic cell i at time t is defined as:

$$\overline{E}_i(t) = \sum_{\vec{N}} E_i P(\vec{N},t) \tag{A1.14}$$

where again the summation extends over all possible socio-configurations.

In the same way man stock variables can be obtained:

$$\overline{M_i^h}(t) = \sum_{\vec{N}} M_i^h P(\vec{N},t) \tag{A1.15}$$

It is now possible to derive equations of motion for the mean values directly from the master equation. For this purpose the master equation is multiplied by $\vec{E}$ and $\vec{M}$ from the left and subsequently summed up via all states $\vec{N}$. However, the equations received subsequently are not yet self-contained since the determination of the right side demands the knowledge of the probability distribution $P(\vec{N},t)$. If it can be assumed that the probability distribution is a well behaved, sharply peaked unimodal distribution

$$\overline{g(\vec{N})} = g(\overline{\vec{N}}) \tag{A1.16}$$

quasi-closed approximate mean value equations can be derived. In this line first the respective process-specific transition rates have to be represented and explained. This is carried out in the following sections for the traffic level and the urban/regional level.

The configurational transition rate $F_t(\vec{E}+\vec{k},\vec{E})$ from population distribution $\vec{E}$ to the neighbouring distribution $\vec{E}+\vec{k}$ is the sum of all the contributions $F_{ij}^{\alpha r}(\vec{E}+\vec{k},\vec{E})$:

$$F_t(\vec{E}+\vec{k},\vec{E}) = \sum_{r=1}^{R}\sum_{\alpha=1}^{P}\sum_{i,j=1}^{L} F_{ij}^{\alpha r}(\vec{E}+\vec{k},\vec{E}) \qquad \text{(A1.17)}$$

where $F_{ij}^{\alpha r}(\vec{E}+\vec{k},\vec{E})$ indicates the number of trips between the traffic cells $i \rightarrow j$ for trip purpose α using traffic mode r. In the total transition rate $F_t(\vec{E}+\vec{k},\vec{E})$ we have neglected contributions referring to simultaneous changes of e.g. trip purpose, traffic mode and location. Since, however, such processes can also be represented by a sequential result of changing, this restriction is not essential for the following steps.

The explicit dependence of the individual terms on $\vec{E}$ indicate that merely those contributions related to a change of the population distribution $\vec{E} \rightarrow \vec{E}+\vec{k}$ are summed up. In this way, a summation via all such terms yields the total transition rate. The index t indicates the possibility of an explicit time dependence. The transition rates have to be specified depending on the kind of interactions/processes. The further procedure is based on the STASA traffic model which has to be extended and generalized for ITEM.

After insertion of the transition rates into the master equation and carrying out of above-mentioned approximations, one receives equations of motion for *the dynamics of the short-run population redistribution:*

$$E_i(t+1) = E_i(t) + \sum_{\alpha}\sum_{r}\sum_{j}\left(F_{ji}^{\alpha r} - F_{ij}^{\alpha r}\right) \qquad \text{(A1.18)}$$

where $E_i(t)$ is the population number of traffic cell i., and t specifies the hour group. Because empirical data are only available for certain hour groups a transition to a difference equation in place of a differential equation is adequate.

The term $F_{ij}^{\alpha r}(\vec{E}+\vec{k},\vec{E})$ describes purpose specific trips from $i \rightarrow j$ by use of the traffic mode r. Therefore, this transition rate corresponds directly to the (empirically countable) traffic flows. Now the question about the driving forces or motives which cause an individual to carry out a trip must be raised.

If, however, panel data are available on the trip-decision behaviour of the different agents of the system (micro-level), the configurational transition rates can directly be computed via:

$$F_{ij}^{\alpha r}(\vec{E}+\vec{k},\vec{E})=\sum_{l\in\Gamma_i} p_{ij}^{\alpha r(l)}(\vec{E},\vec{M},\vec{\kappa}_l), \tag{A1.19}$$

where one has to sum up over all individual contributions of all agents belonging to subpopulation α and travelling from traffic cell i to traffic cell j using mode r. This procedure is however very extensive because of the immensity of the required data base. Therefore it can only be carried out successfully in exceptional situations. Consequently, a less exhaustive procedure which is more amenable to the analysis and the simulation work is indicated.

Since $E_i(t)$ individuals are at time t in area i one can define an „individual" transition rate $p_{ij}^{\alpha r}(\vec{E},\vec{M},\vec{\kappa})$. The number of the trips $F_{ij}^{\alpha r}(\vec{E}+\vec{k},\vec{E})$ is therefore given by

$$F_{ij}^{\alpha r}(\vec{E}+\vec{k},\vec{E})=E_i p_{ij}^{\alpha r}(\vec{E},\vec{M},\vec{\kappa}) \tag{A1.20}$$

for $\vec{k}=\left\{0,\ldots,1_j^{\alpha r},\ldots,0,\ldots,(-1)_i^{\alpha r},\ldots,0,\ldots\right\}$, which describes a trip from i to j, and $F_{ij}^{\alpha r}(\vec{E}+\vec{k},\vec{E})=0$ for all other $\vec{k}$.

Since $E_i(t)$ agents are at time t in the traffic cell i, the probability for a trip to another traffic cell is proportional $E_i(t)$. Let $p_{ij}^{\alpha}(\vec{E},\vec{x})$ be the transition rate from i to j for trip purpose α. Of course, this transition rate depends among other things on the explicit distribution of the agents $\vec{E}$ and cell specific characteristics $\vec{x}$ of the infrastructure, for example, job supply, the housing market, services available for companies and households as well as leisure facilities, to mention a few (HAAG, 1990). In this way, the number of the trips between i and j is given by:

$$F_{ij}^{\alpha}(t) = E_i(t)p_{ij}^{\alpha}(\vec{E},\vec{x},t), \tag{A1.21}$$

where $\vec{k} = \left\{0,\ldots,1_j^{\alpha r},\ldots,0,\ldots,(-1)_i^{\alpha r},\ldots,0,\ldots\right\}$ represent trip probabilities from one traffic cell i into j, and $F_{ij}^{\alpha r}(\vec{E}+\vec{k},\vec{E}) = 0$ for all other $\vec{k}$.

The dynamics of the (mean) population redistribution on the macro level can be derived directly by an averaging procedure from the master equation:

$$\frac{dE_i(t)}{dt} = \sum_{\alpha}\sum_{j=1}^{LMAX} E_j(t)p_{ji}^{\alpha}(\vec{E},\vec{x}) - \sum_{\alpha}\sum_{j=1}^{LMAX} E_i(t)p_{ij}^{\alpha}(\vec{E},\vec{x}) \tag{A1.22}$$

The change of the population number $E_i(t)$ of a traffic cell i of one hour to the next hour can be calculated on the basis of the traffic flows F_{ij} and F_{ji} between the different traffic cells i and j. The flows themselves in turn depend on the respective attractivities $u_i^{\alpha}(\vec{E},\vec{x})$ of the traffic cells, the resistance functions $g^{\alpha}(w_{ij})$ and the scaling parameter (mobility parameter) $\varepsilon^{\alpha}(t)$. In this way, the population distribution and therefore the car (vehicle) distribution are represented in a time-dependent manner. This is required in order to count traffic demand and traffic supply in an appropriate manner. In particular the determination and estimation of daily traffic peaks can only be carried out using a time-dependent transport model.

The individual transition rates $p_{ij}^{\alpha r}(\vec{E},\vec{M},\vec{\kappa})$ represent group related trip probabilities from one traffic cell to another one, where essentially three factor sets are important:

- attractivities $u_i^{\alpha}(\vec{E},\vec{x})$ of the particular traffic cells which depend across-the-board on e.g. the population distribution, the distribution of the work places and dwellings. The significance of the different socio-economic variables $\vec{x}$ is determined by means of a multivariate regression procedure. Obviously the composition of the particular set of key (significant) variables depends strongly on the trip purpose α.

- resistance's $g^{\alpha}(w_{ij})$ depending on trip purpose α, where w_{ij} represents a generalised resistance parameter $w_{ij}^{\alpha r} = t_{ij}^{\alpha r} + b_1^{\alpha} c_{ij}^{r} + b_2^{\alpha} K_{ij}^{\alpha r}$ for a trip from *i* to *j*, where t_{ij} is the traffic density dependent travel time t_{ij}, c_{ij}^{r} the travel costs, $K_{ij}^{\alpha r}$ appropriate comfort parameters, and $b_1^{\alpha}, b_2^{\alpha}$ are the corresponding weight factors. The following resistance function seems to be adequate in order to describe modal choice processes (second term) as well as changes in the time of travel (third term):

$$g^{\alpha}(w_{ij}^{\alpha r}(\tau)) = c_0^{\alpha r}(w_{ij}^{\alpha r}(\tau))^{c_1^{\alpha r}} \frac{\exp(-c_2^{\alpha r} w_{ij}^{\alpha r}(\tau))}{\sum_{s=1}^{R} \exp(-c_2^{\alpha s} w_{ij}^{\alpha s}(\tau))} \frac{\exp(-c_3^{\alpha r} w_{ij}^{\alpha r}(\tau))}{\sum_{\tau''=\tau-1}^{\tau+1} \exp(-c_3^{\alpha r} w_{ij}^{\alpha r}(\tau'))} \qquad \text{(A1.23)}$$

In a first step only travel times t_{ij} are considered in the resistance parameter w_{ij} for data reasons. In accordance with HAUTZINGER (1982) and STEIERWALD SCHOENHARTING (1993a), the following simplified resistance function of the gamma type seems to be adequate:

$$g^{\alpha}(t_{ij}) = t_{ij}^{c_1^{\alpha}} \exp(-c_2^{\alpha} t_{ij}), \qquad \text{(A1.24)}$$

where the parameters $c_0^{\alpha r}, c_1^{\alpha r}, c_2^{\alpha r}$ and $c_3^{\alpha r}$ depend on trip purpose α, the travel mode r and the specific trip hour and day, respectively.

- and a time-dependent scaling parameter $\varepsilon^{\alpha}(t)$ that correlates with the mobility behaviour.

For the traffic flows of the STASA transport model, the following functional dependence proved effective:

$$F_{ij}^{\alpha r} = E_i p_{ij}^{\alpha r}(\vec{E}, \vec{M}) = E_i \nu^{\alpha r}(t) f^{\alpha r} b_t g\ (w_{ij}^{\alpha r}) \exp\left[\gamma^{\alpha} u_j^{\alpha}(\vec{E}, \vec{M}) - u_i^{\alpha}(\vec{E}, \vec{M})\right] \qquad \text{(A1.25)}$$

where the flexibility of the agents to undertake a trip is considered by the scaling parameter $\varepsilon^{\alpha}(t)$, the width of the hour group means by b_t and the parameter $f^{\alpha r}$ indicates the accessibility of traffic mode *r*.

The described modelling of the traffic flows is based on the general experience that agents compare the attractivities of the traffic cells and that the probability for a trip $i \rightarrow j$ increases with increasing differences $\left[\gamma^{\alpha} u_j^{\alpha}(\vec{E},\vec{M}) - u_i^{\alpha}(\vec{E},\vec{M})\right] > 0$ of attractivities per unit of time and simultaneously the trip probability into the opposite direction $j \rightarrow i$ decreases (assuming the same resistance function in both directions). In this sense the attractivity differences are the "driving forces" of the transport/traffic system. The different weighting of the source cell and the target cell, and therefore the differently strong influence of the site factors on the generation and distribution of traffic is considered by the parameter γ^{α}.

The macroscopic traffic flows are usually different even for identical attractivities of the traffic cells, since they have to be multiplied by the population numbers E_i and E_j, respectively.

On the other hand, a simultaneous increase of the attractivities of all traffic cells by a common factor may not change the trip distribution. This is guaranteed by the normalization of the attractivities

$$\sum_{i=1}^{L} u_i^{\alpha}(\vec{E},\vec{M}) = const. \tag{A1.26}$$

The transportation system is in its (macroscopic) balance, if simultaneously for all traffic cells the sum of inflows into a traffic cell equals the sum of outflows.

The internal traffic within a traffic cell is given by:

$$F_{ii}^{\alpha r} = E_i p_{ii}^{\alpha r}(\vec{E},\vec{M}) = E_i \nu^{\alpha r}(t) f^{\alpha r} b_t g\ (w_{ii}^{\alpha r}) \exp\left[(\gamma^{\alpha} - 1) u_i^{\alpha}(\vec{E},\vec{M})\right] \tag{A1.27}$$

As expected, the internal traffic of a cell *i* depends directly on the corresponding socio-economic data of that particular cell and via the normalization of the attractivities on the data of all other cells as well.

However, up to now the portrayed formalism describes only *short-term effects* like redistribution of the population (A1.22) and the internal traffic flows (A1.27). *Long-term effects* e.g. migration of the population between cells as well as structural development effects have to be considered too.

The dependence of the attractivities $u_i^\alpha(\vec{E},\vec{x})$ of the traffic cells on the population distribution $\vec{E}$ and further socio-economic variables $\vec{x}$ has the consequence that the evolution equation now becomes a non-linear differential equation or difference equation. In this case, non-linearities among other things reflect the complexity of the decision processes involved. Depending on the initial conditions i.e. the agents and/or vehicle distribution at a given time and the further system parameters of the urban/regional areas, the non-linear dynamics can lead to a complex variety of self-organising traffic flows. Consequently, the dynamics of the population distribution is described by traffic flows and feedback effects resulting from the land-use patterns depending on the time of day.

The total number of the traffic cells is 94. Since there exist 6 main traffic directions *LMAX* is 94+6=100 for the region of Nanjing. In this way, through traffic as well as source and target traffic are considered as well.

In total the STASA transport model contains trip-chains in aggregated form. Here a decisive difference appears to traditional traffic models, since the hourly (day time-controlled) redistribution of the agents can be considered and simulated. More precisely, the so called four-stage-algorithm is not able to describe trip-chains as well as time-of-day dependent effects. Because the decisions concerning the transport system depend on the respective current distribution of the agents, that is from the particular number of the agents in the respective traffic cells, a dynamic redistribution of the population is not only plausible it improves simultaneously the model quality. It is a further advantage of the STASA model that traffic generation and traffic redistribution occurs at one model step. In detail, the determination of the utility functions and mobilities of the migrating population is portrayed in section 7.2.

The STASA-approach allows - unlike other traffic models - the description of the traffic generation and traffic distribution in one step. The urban or regional socio-economic factors influence as push/pull-factors the attractivities $u_i^\alpha(\vec{E},\vec{M})$ and $u_j^\alpha(\vec{E},\vec{M})$ and are therefore simultaneously responsible for traffic generation and distribution.

This leads to the formulation of the following central questions:

- Which factors determine the general frequency to change the target area?
- Which factors refer to hindrances or barriers to rides?

- Which „key factors" determine the different attractivities of the individual traffic cells?

These three questions will be answered in the following sections.

The trip-purpose specific attractivities of the traffic cells

The wanted attractivities of the respective traffic cells are gained from two different fields:

- Synergy variables are describing on the one hand a general group behaviour in the form of pigeon and/or band-waggon effects and on the other hand saturation effects and/or negative externalities.

- Trip-purpose specific attractivity factors depending on time of day can be appropriate and are related to very different factor sets. Their relevance must be determined via a procedure of multiple regression.

The selection of variables has to take into account the mentioned mutual dependence between the micro- and macro-level and is based empirically on experiences like the following:

Traffic flows between cells may exhibit a self-acceleration process. This process does not however increase without limits but depends on specific saturation boundaries. For the latter the accessibility of traffic cells referring to the traffic modes as well as the corresponding travel times and costs are essential.

and furthermore:

An observed traffic cell seems to be more attractive the more concentrated specific trip-purpose related functions or characteristics are and the higher their social stability is.

Parameter estimation

The system parameters, such as the mobilities, the attractivities and the parameters of the resistance function occurring in the approach for the trip frequency can be directly linked to the trip decision processes by different optimization procedures. For this aim the ITEM transport model has to be matched to the empirical traffic flow matrices (index e), the population numbers and travel time matrices, which in turn are dependent on the traffic volume of the hour groups, respectively

$$F_{ij}^{\alpha e}(\tau), E_i^e(\tau), t_{ij}(F_{ij}^{\alpha e}), \qquad \text{(A1.28)}$$

for the hour groups under consideration at time $\tau = 1,...,4$. The minimization of the functional

$$H_t\left[\varepsilon^\alpha, c_1^\alpha, c_2^\alpha, \vec{u}^\alpha\right] = \sum_{i,j=1}^{L}\left[F_{ij}^{e\alpha}(\tau) - \varepsilon^\alpha(\tau)b(\tau)E_i^e(\tau)g^\alpha(w_{ij}(\tau))\exp(u_j^\alpha(\tau) - u_i^\alpha(\tau))\right]^2 = Min.$$

(A1.29)

enables us to calculate the optimal attractivities $u_i^\alpha(\vec{E},\vec{x})$ and further system parameters for each trip purpose and each time of day. Depending on the time of day, the trip purpose distribution changes within the (macroscopic) traffic flows. For example, trips between home - work and work - home may predominate in the morning hours and the late afternoon , while leisure time trips or shopping trips show another temporal distribution over the day. Therefore, the estimated parameters are dependent on the trip purpose and time of day.

The wanted site factors $x_i^n, n = 1,...$ of the traffic cells are gained from two different fields: on the one hand from the class of the so-called synergy variables, describing general group effects such as pigeon effects and/or. band-wagon effects and on the other hand from a sequence of potential attractivity factors e.g. number of jobs available, the number of vacant dwellings, regional income per capita, and other local infrastructure dependent factors. The set of key-factors (selected socio-economic variables) is in a second step determined via a multiple regression analysis:

$$u_i^\alpha(\vec{E},\vec{x}) = \sum_n b_n^\alpha(\tau)x_i^n(\tau) \qquad \text{(A1.30)}$$

The elasticity's $b_n^\alpha(\tau)$ assigned to the socio-economic variables $x_i^n(\tau)$ are dimensionless numbers and indicate the influence of the independent variables on the dependent variable. The selection of relevant site factors occurs by means of the corresponding statistical characteristics (T-values, other significance tests).

Appendix A2: Maps of the Urban Area, Nanjing City and Greater Nanjing

Figures A2: Urban Area, Nanjing City, and Greater Nanjing

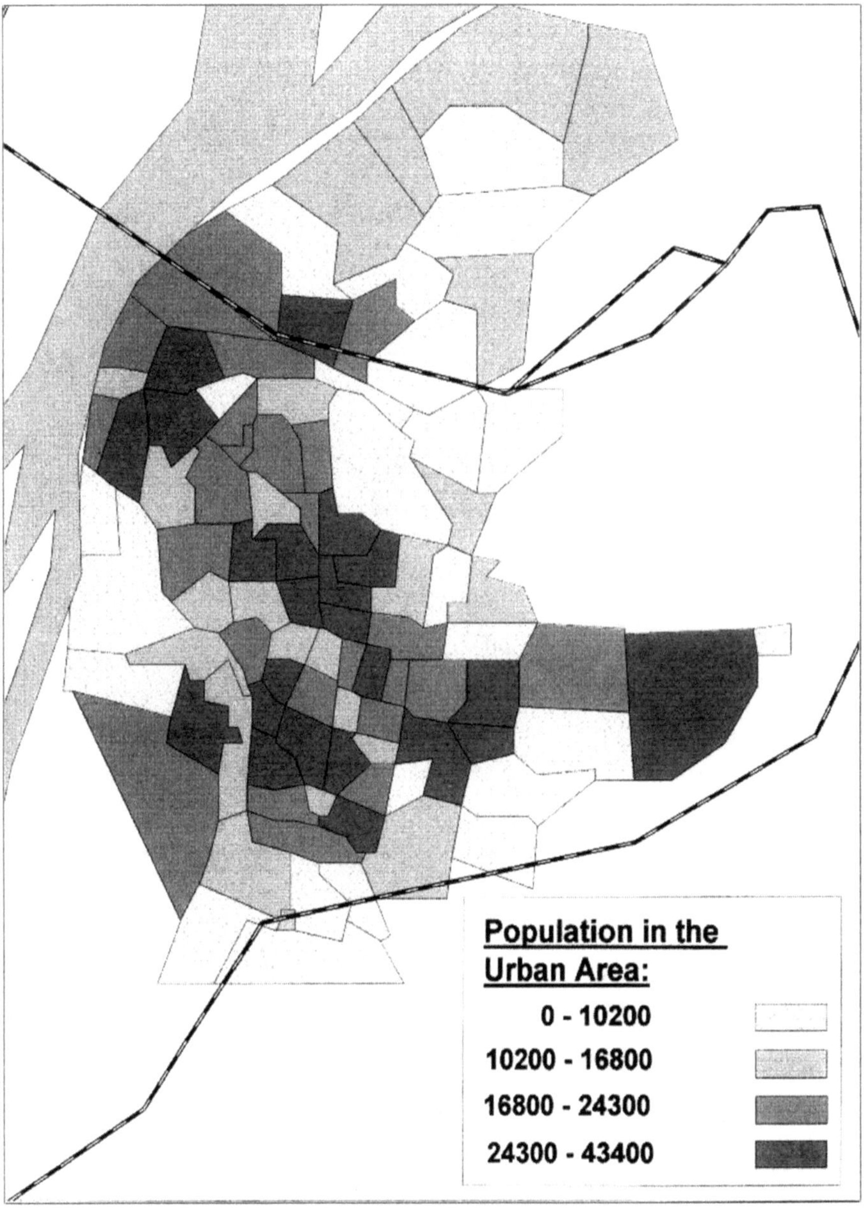

Figure A2.1 The *Urban Area* of Nanjing with population numbers

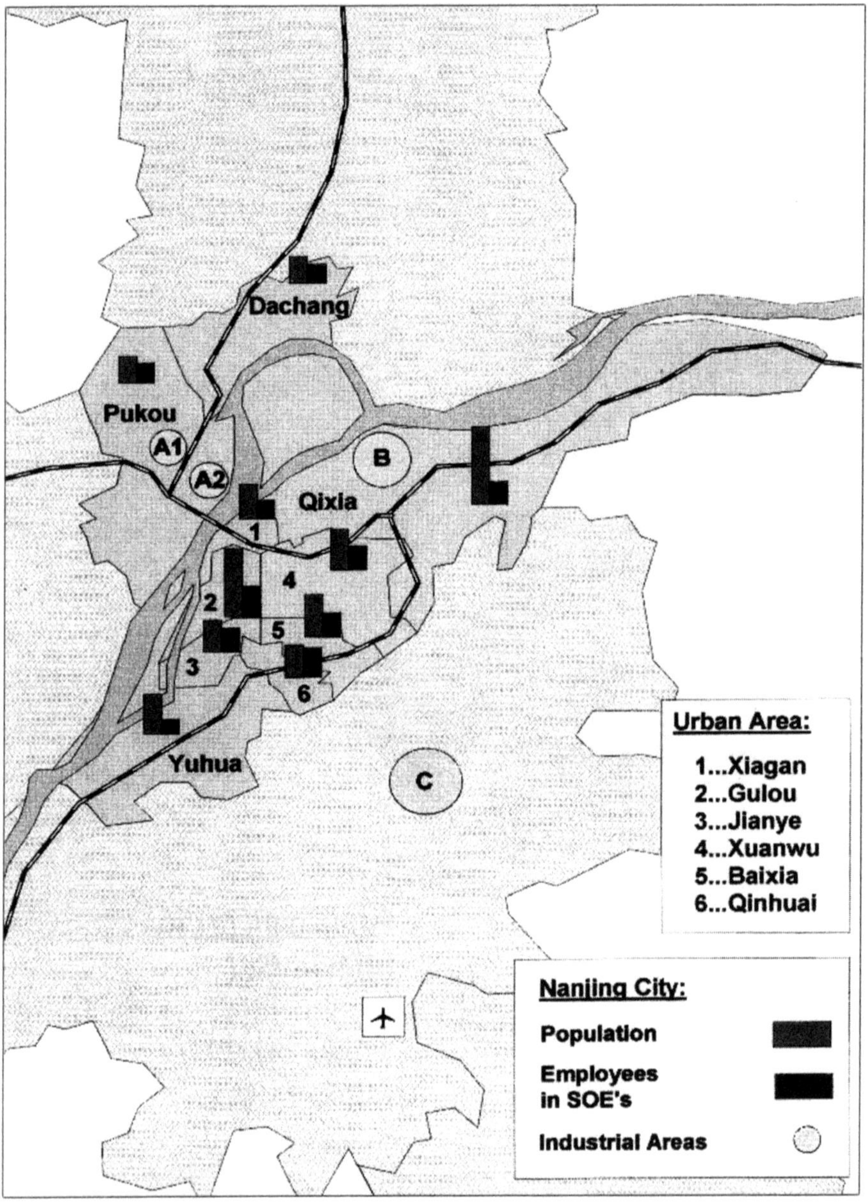

Figure A2.2 **Map of *Nanjing City* with population numbers, employees in SOE's and industrial areas**

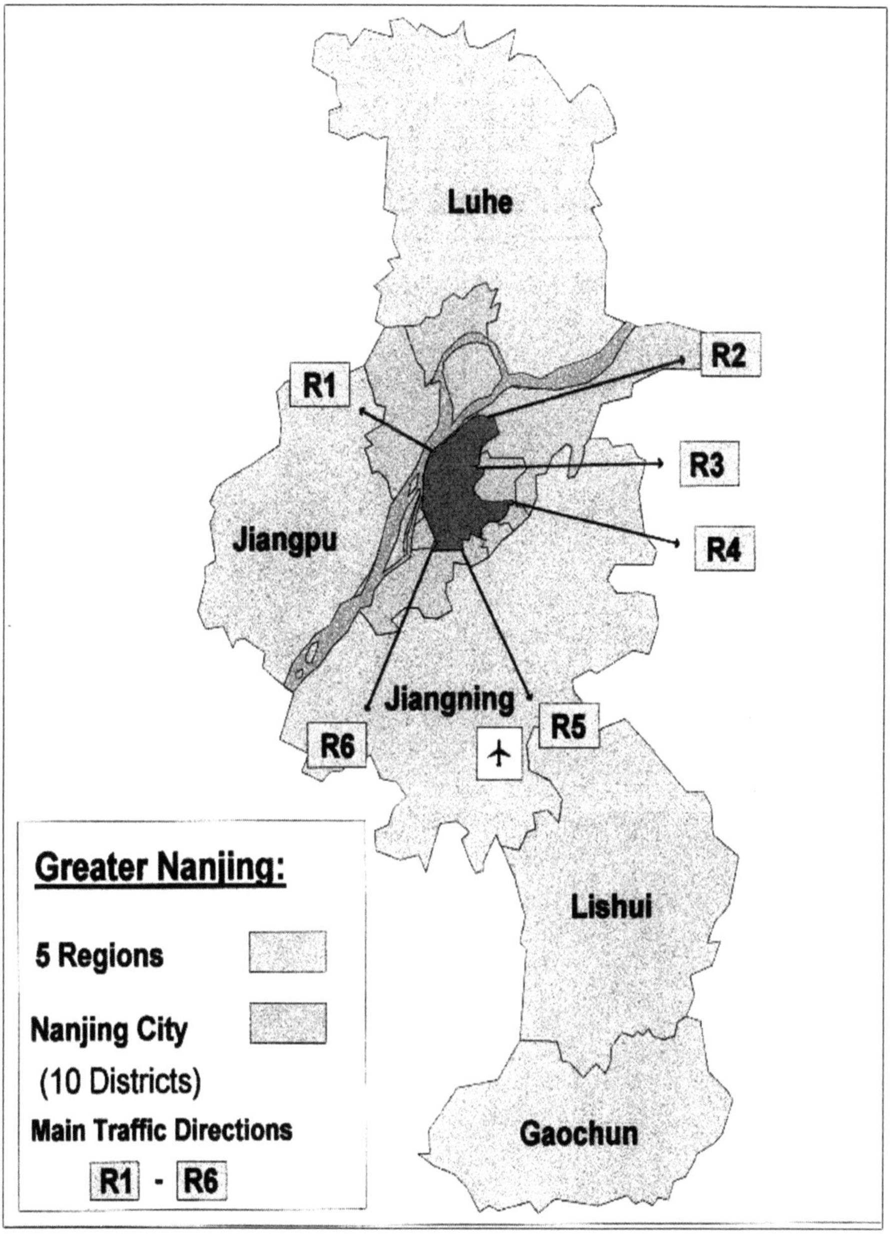

Figure A2.3 **Map of *Greater Nanjing* with main traffic directions and airport**

Appendix A3: General Figures

Figures A3.1: Attractivities

Fig. A3.1a: Attractivities with respect to the trip purpose *home-work* of 1987 in the Urban Area

Fig. A3.1b: Attractivities with respect to the trip purpose *home-education* of 1987 in the Urban Area

Fig. A3.1c: Attractivities with respect to the trip purpose *shopping* of 1987 in the Urban Area

Figures A3.2: Population numbers

Fig. A3.2a: Real population numbers in the Urban Area of *1987*

Fig. A3.2b: Real population numbers in the Urban Area of *1996*

Fig. A3.2c: Calculated population numbers in the Urban Area of *2010*

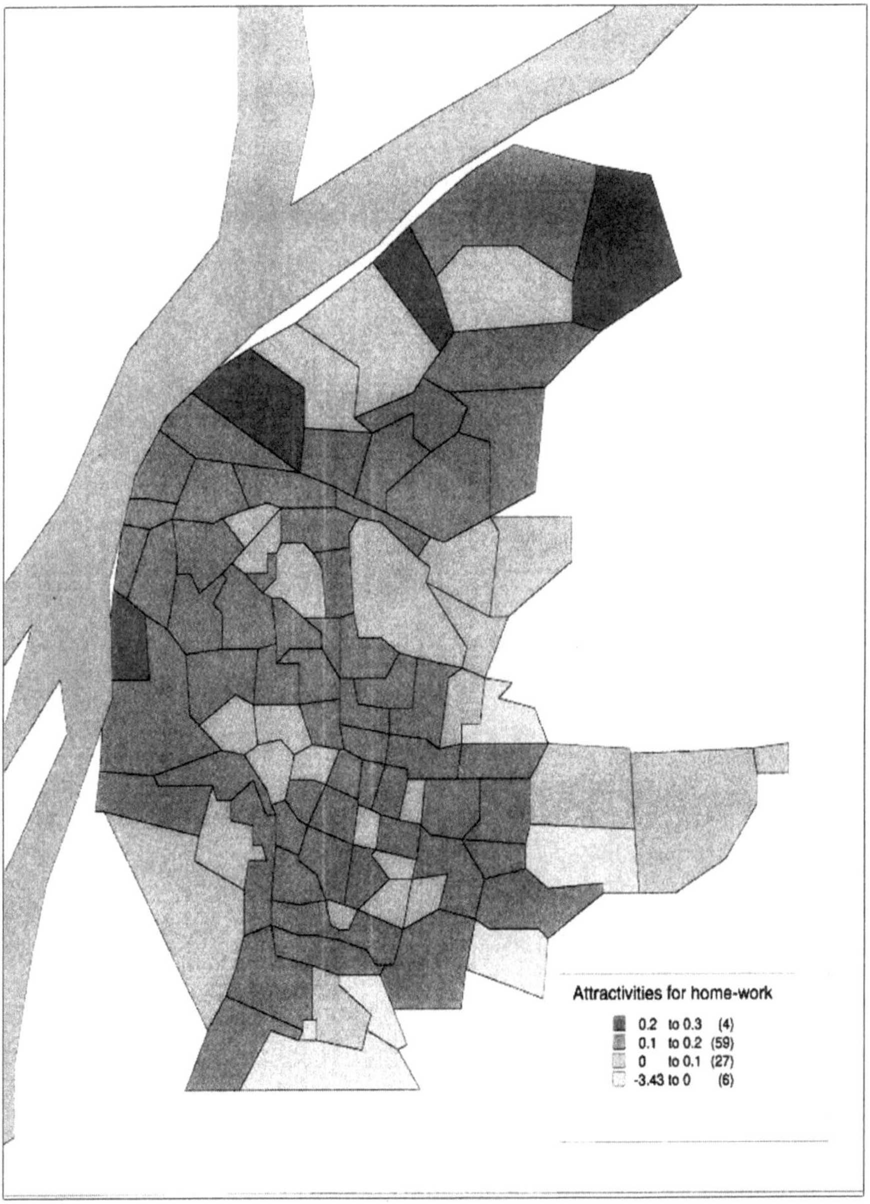

Figure A3.1a Attractivities of the cells in the Urban Area with respect to the trip purpose *home – work*

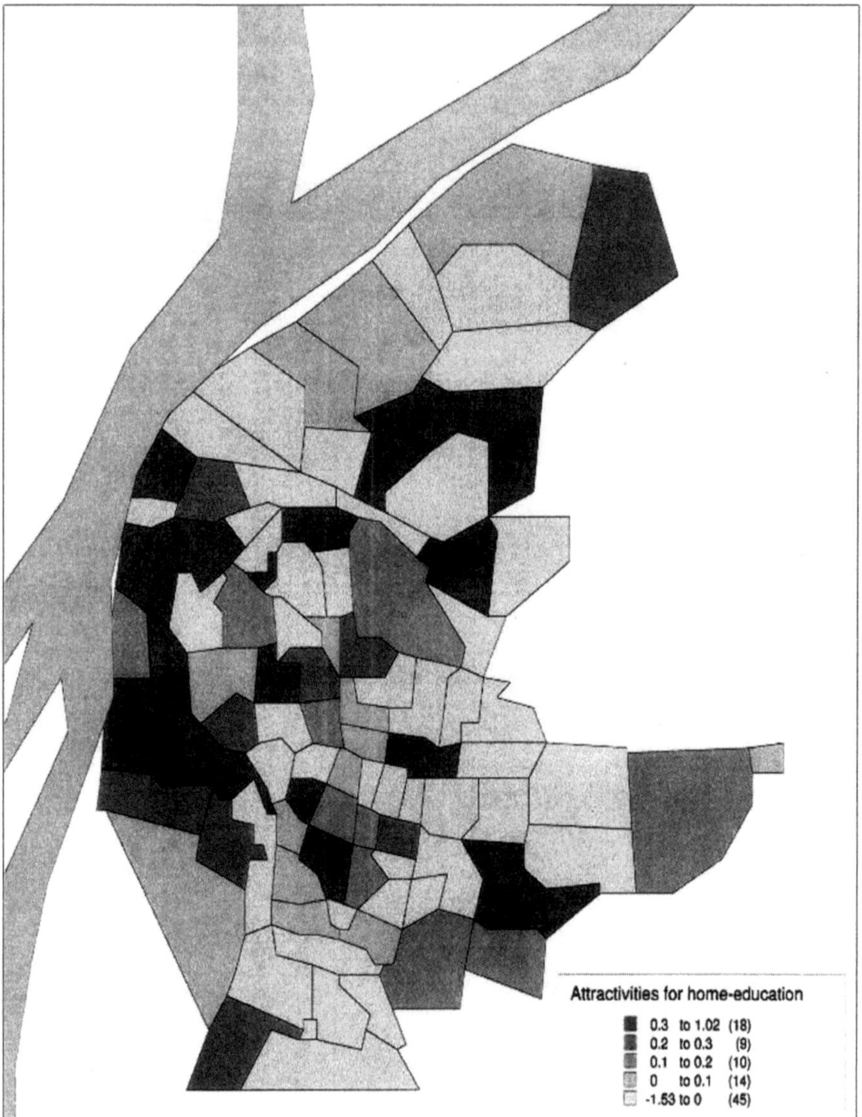

Figure A3.1b Attractivities of the cells in the Urban Area with respect to the trip purpose *home - education*

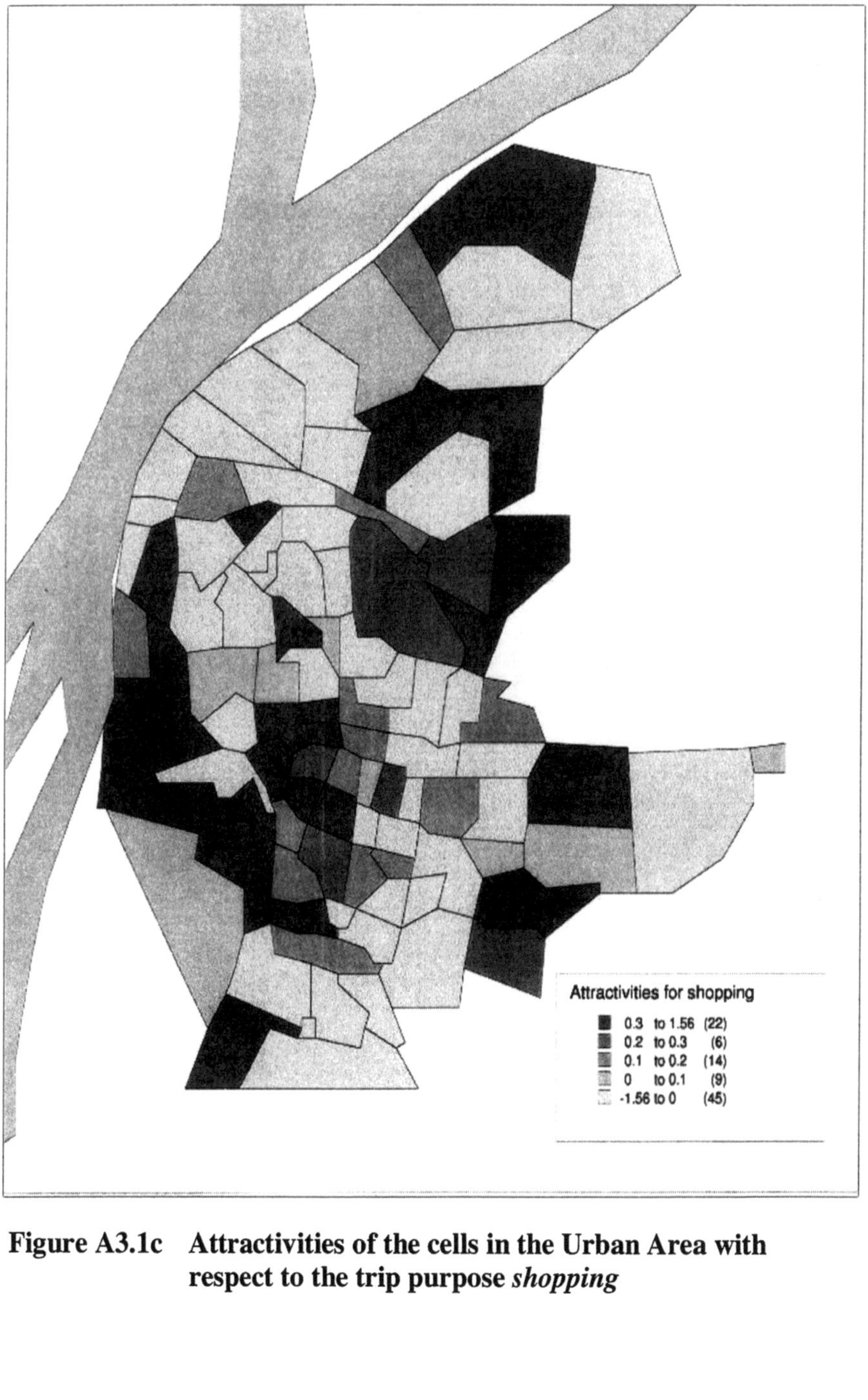

Figure A3.1c Attractivities of the cells in the Urban Area with respect to the trip purpose *shopping*

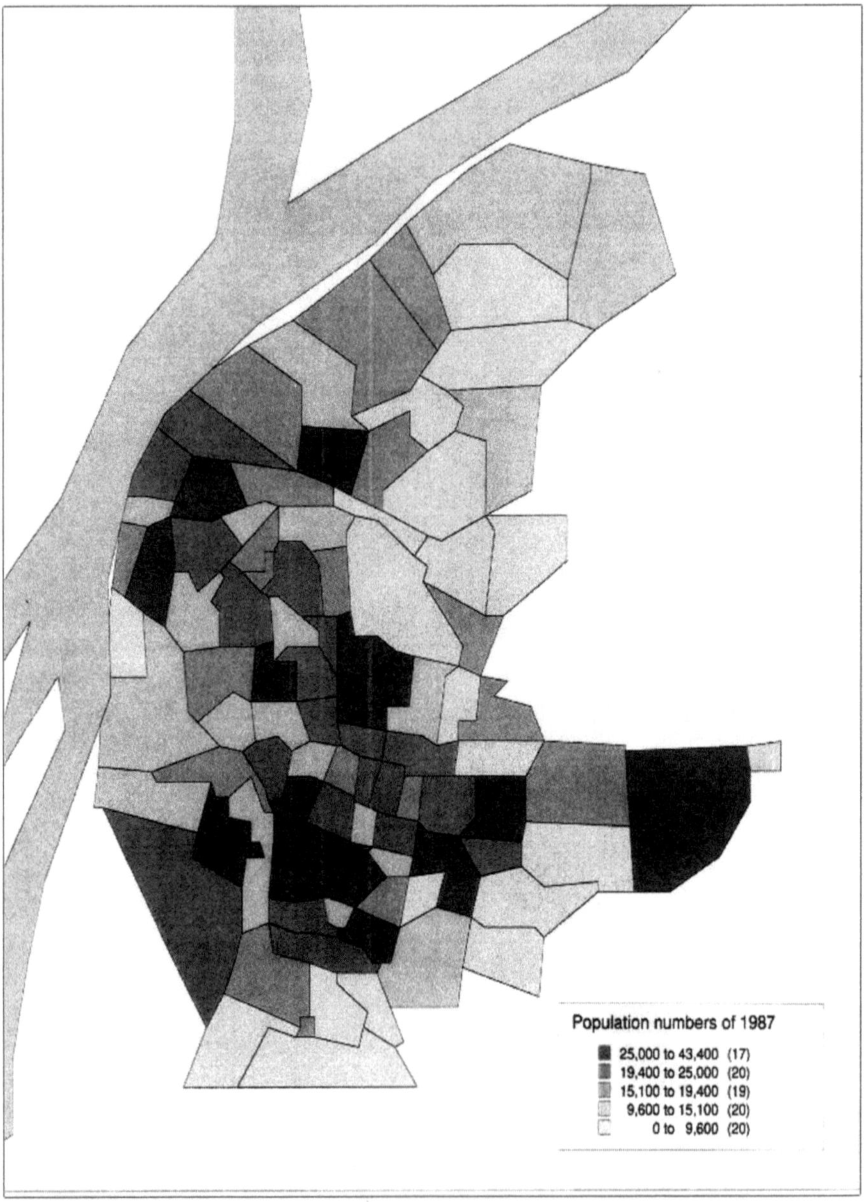

Figure A3.2a Population numbers in the Urban Area of *1987*

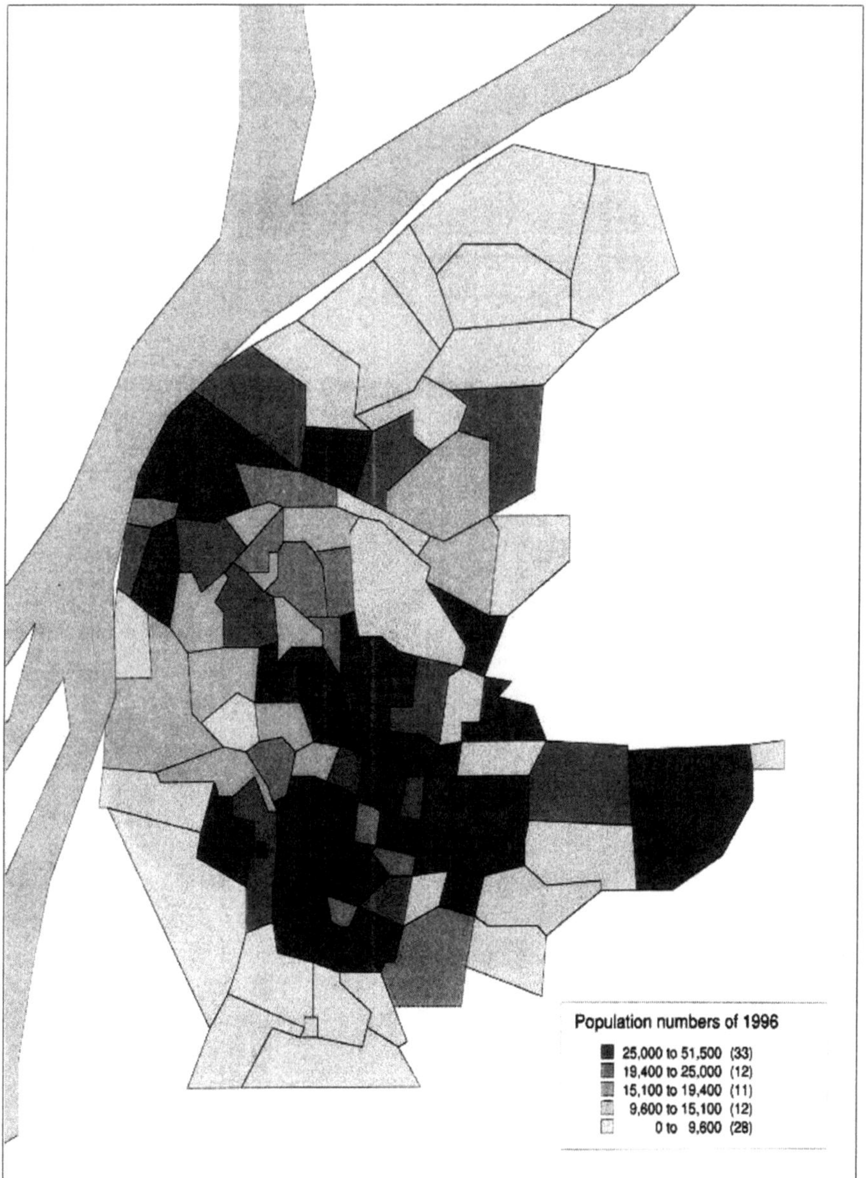

Figure A3.2b Population numbers in the Urban Area of *1996*

Figure A3.2c Calculated population numbers in the Urban Area of *2010*

Appendix A4: Figures of the Scenarios

Figures A4.1 – A4.5:	**Existing areas of 1990 and planned areas of 2010 with respect to the four categories *industry*, *housing*, *business*, and *greenbelt***
Fig.A4.1a:	Existing areas of 1990: *industry*
Fig.A4.1b:	Planned areas of 2010: *industry*
Fig.A4.2a:	Existing areas of 1990: *housing*
Fig.A4.2b:	Planned areas of 2010: *housing*
Fig.A4.3a:	Existing areas of 1990: *business*
Fig.A4.3b:	Planned areas of 2010: *business*
Fig.A4.4a:	Existing areas of 1990: *greenbelt*
Fig.A4.4b:	Planned areas of 2010: *greenbelt*

Figures A4.5 – A4.8:	**Network assignment, average speed (source), source traffic, target traffic and population numbers of the different scenarios**
Fig.A4.5a:	Basic scenario E: *network assignment*
Fig.A4.5b:	Basic scenario E: *average speed* (source)
Fig.A4.5c:	Basic scenario E: *source traffic*
Fig.A4.5d:	Basic scenario E: *target traffic*

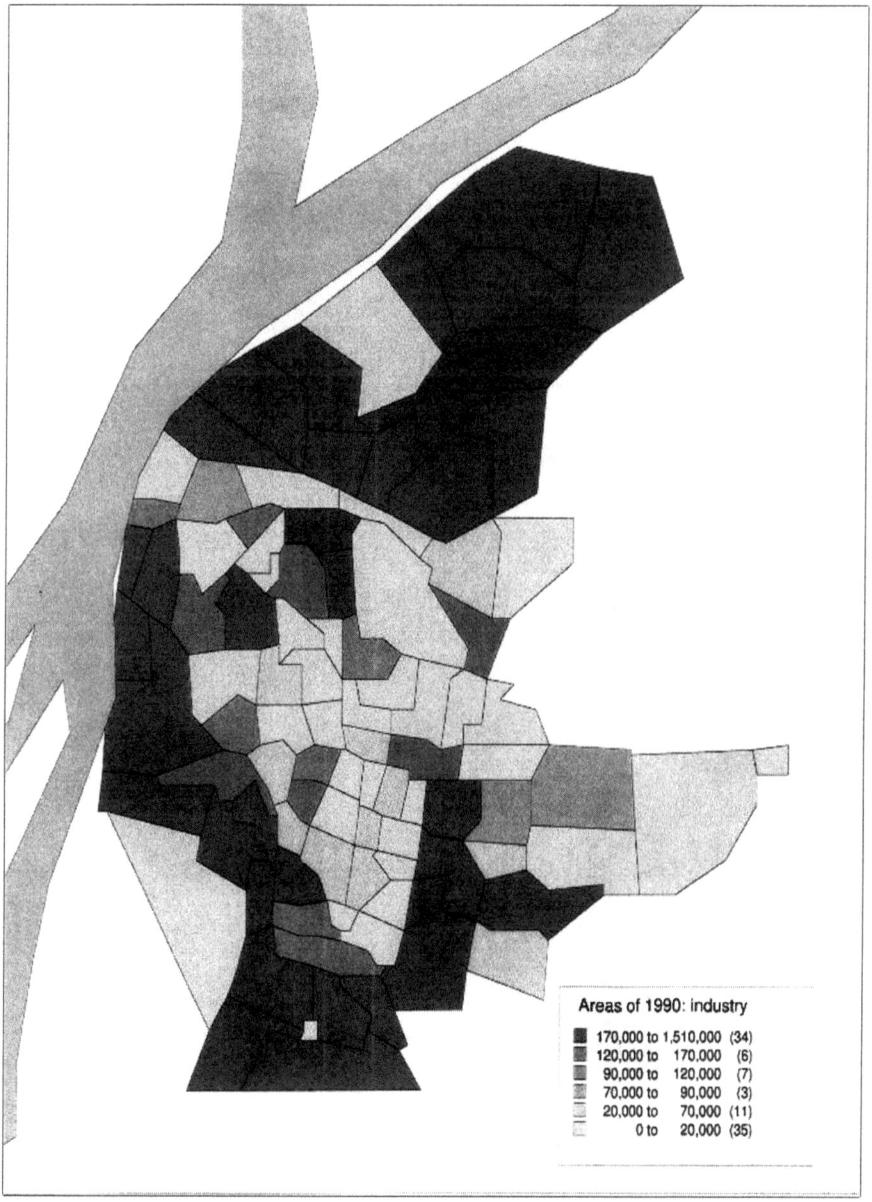

Figure A4.1a Square meters with respect to the Category *industry* in the Urban Area for the year 1990

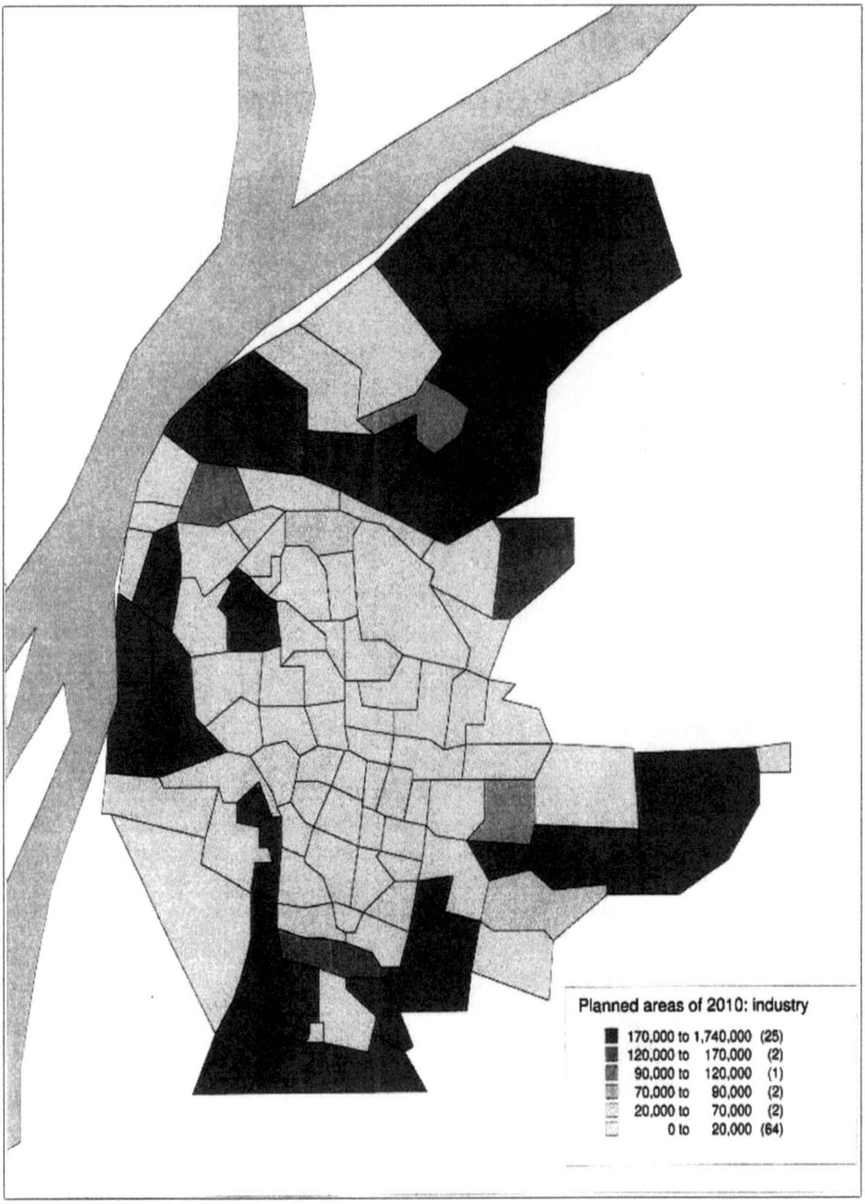

Figure A4.1b Square meters with respect to the Category *industry* in the Urban Area for the year 2010

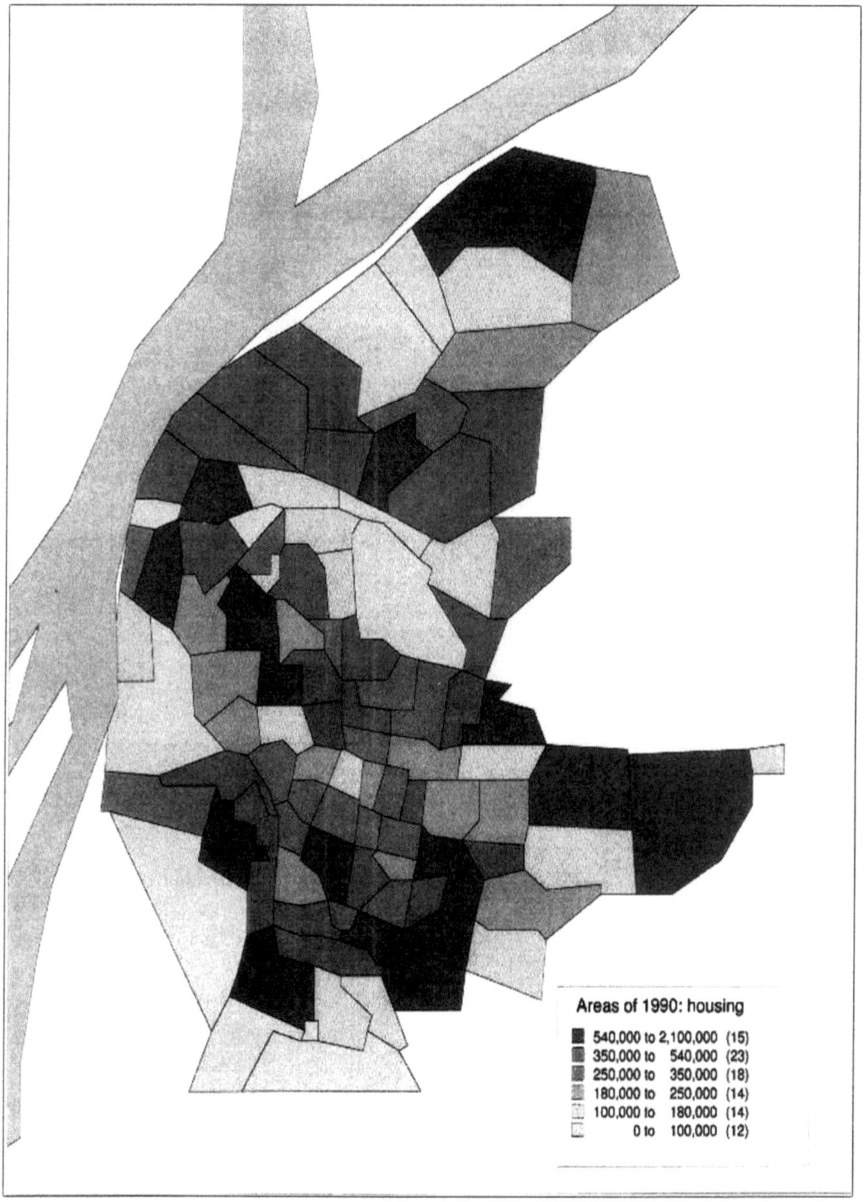

Figure A4.2a Square meters with respect to the Category *housing* in the Urban Area for the year 1990

Figure A4.2b Square meters with respect to the Category *housing* in the Urban Area for the year 2010

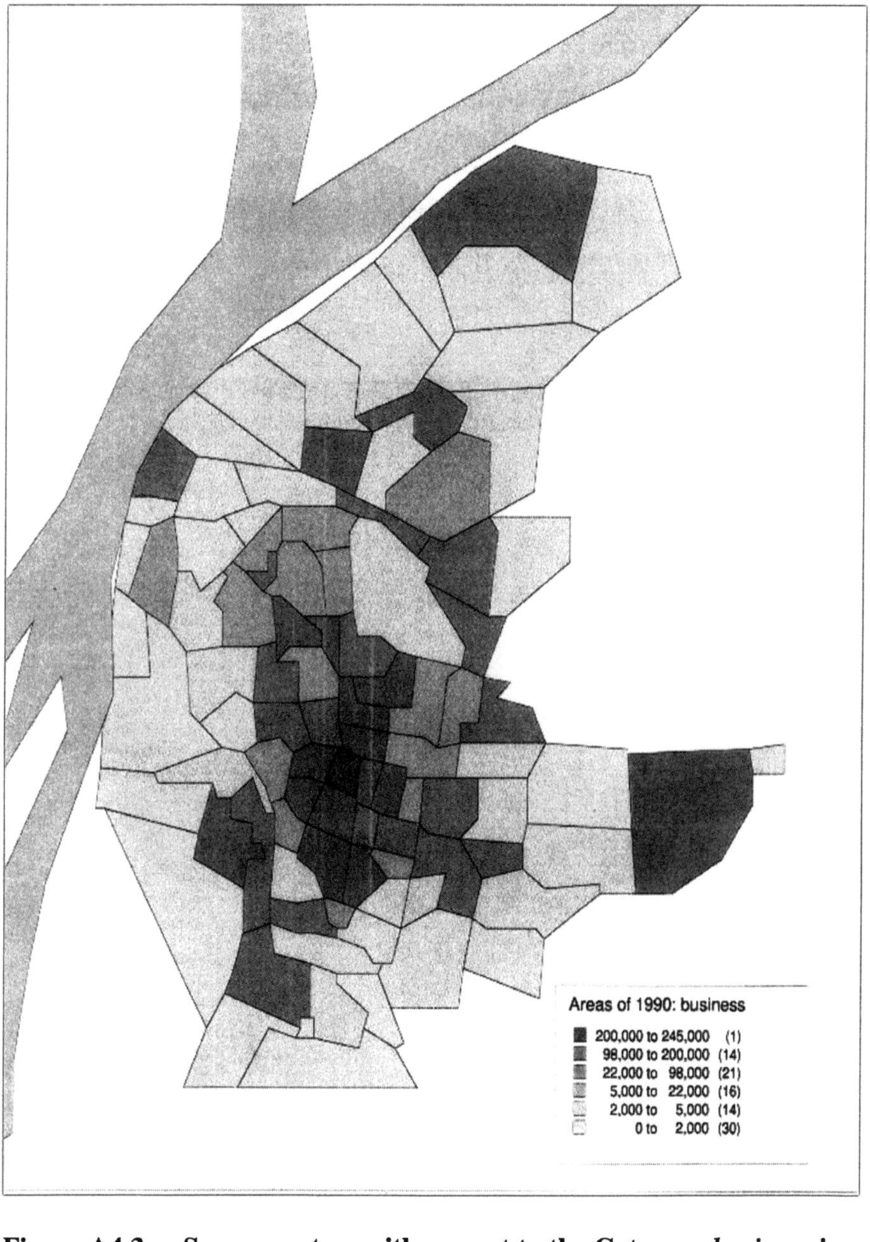

Figure A4.3a Square meters with respect to the Category *business* in the Urban Area for the year 1990

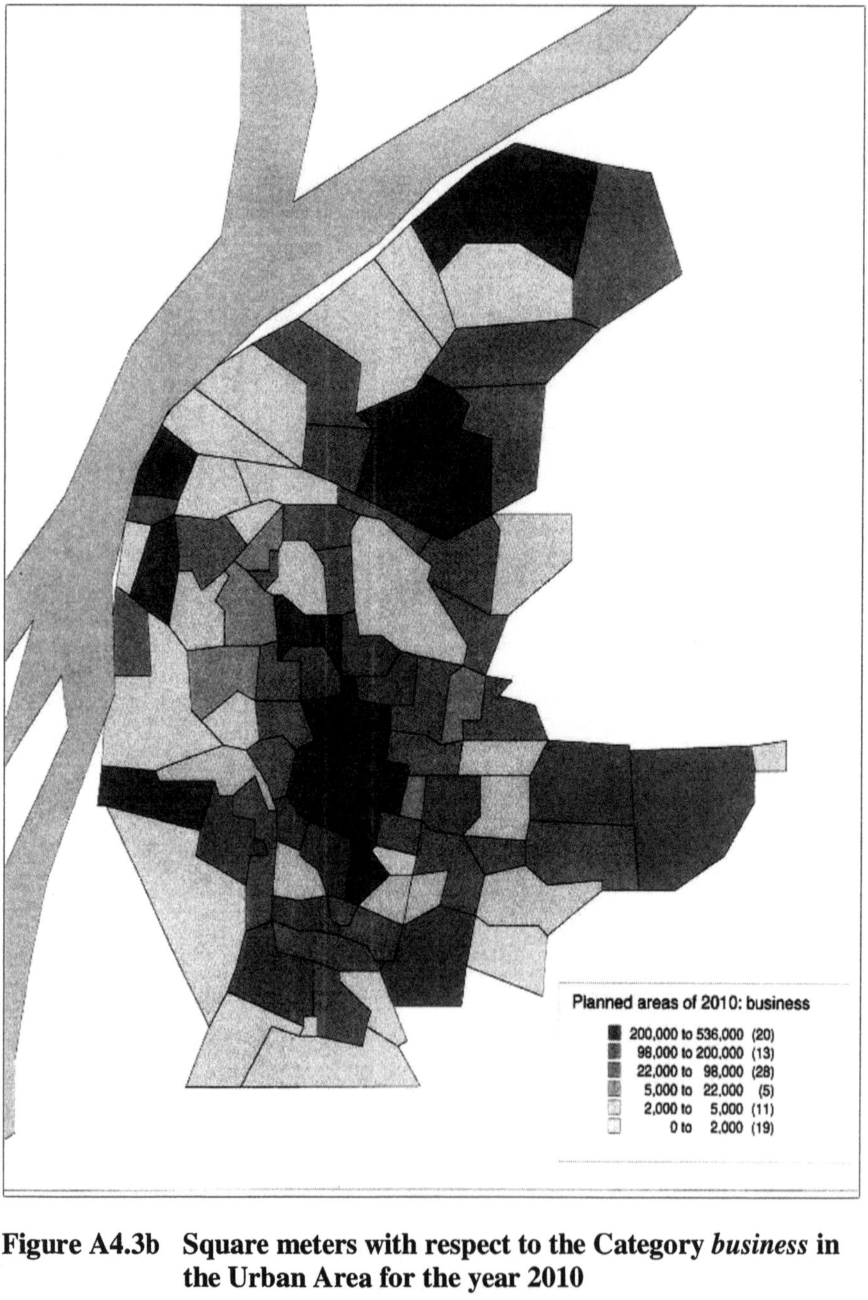

Figure A4.3b Square meters with respect to the Category *business* in the Urban Area for the year 2010

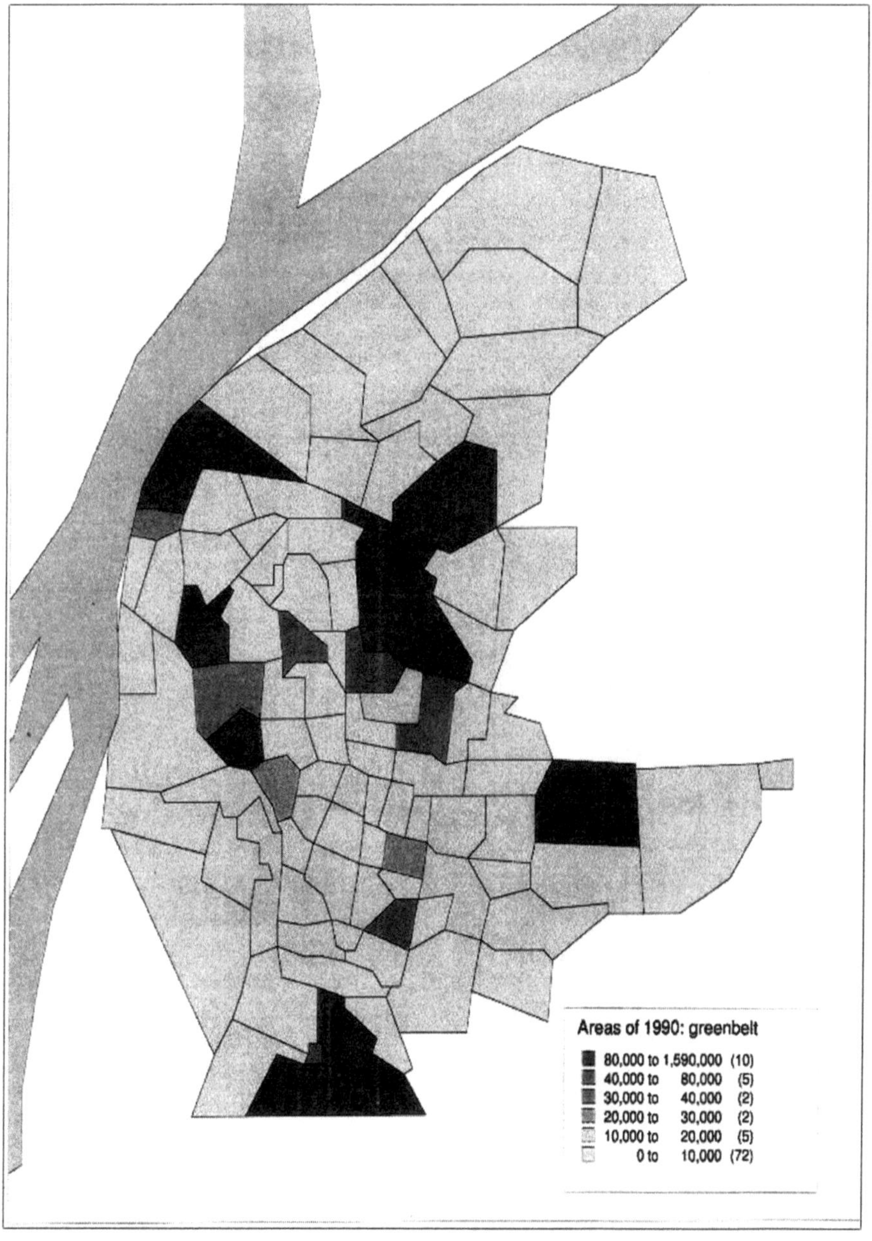

Figure A4.4a Square meters with respect to the Category *greenbelt* in the Urban Area for the year 1990

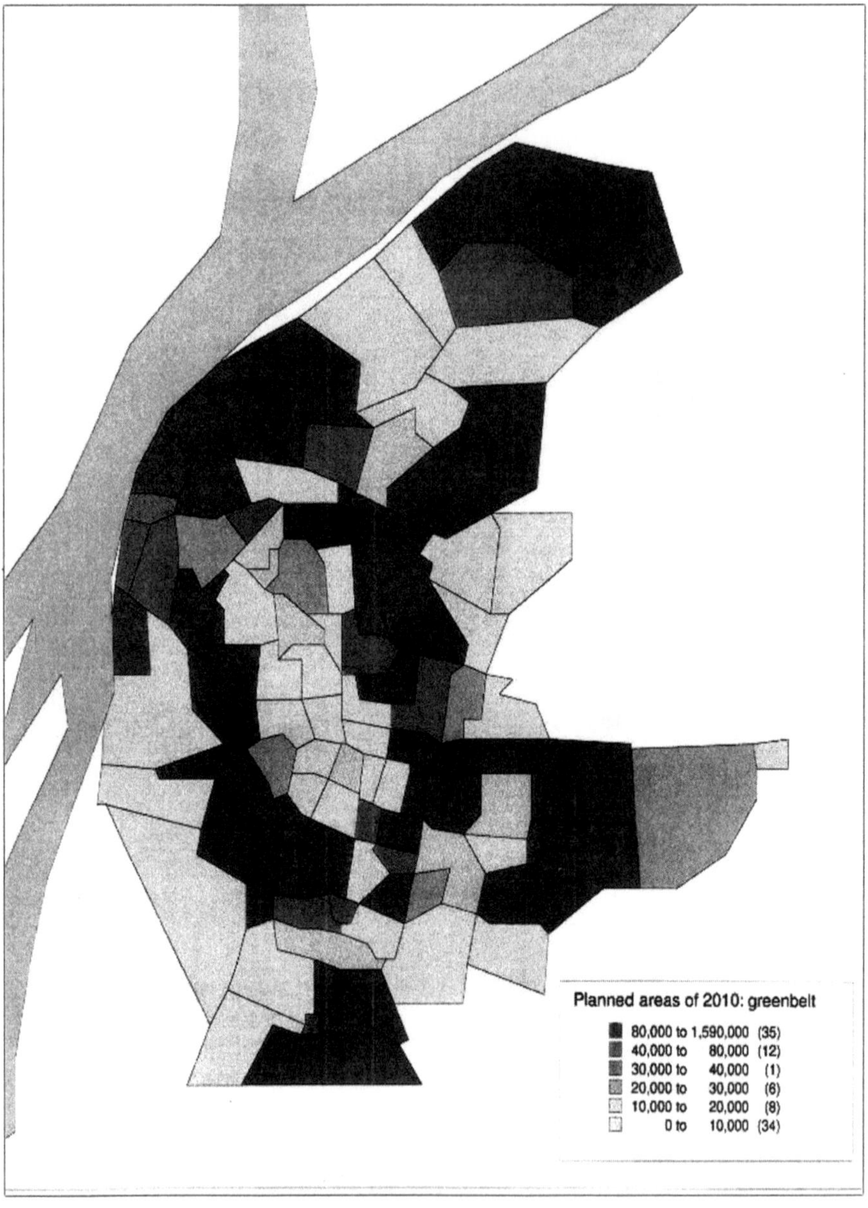

Figure A4.4b Square meters with respect to the Category *greenbelt* in the Urban Area for the year 2010

Figure A4.5a Basic scenario: *Assignment* of the street network in the Urban Area

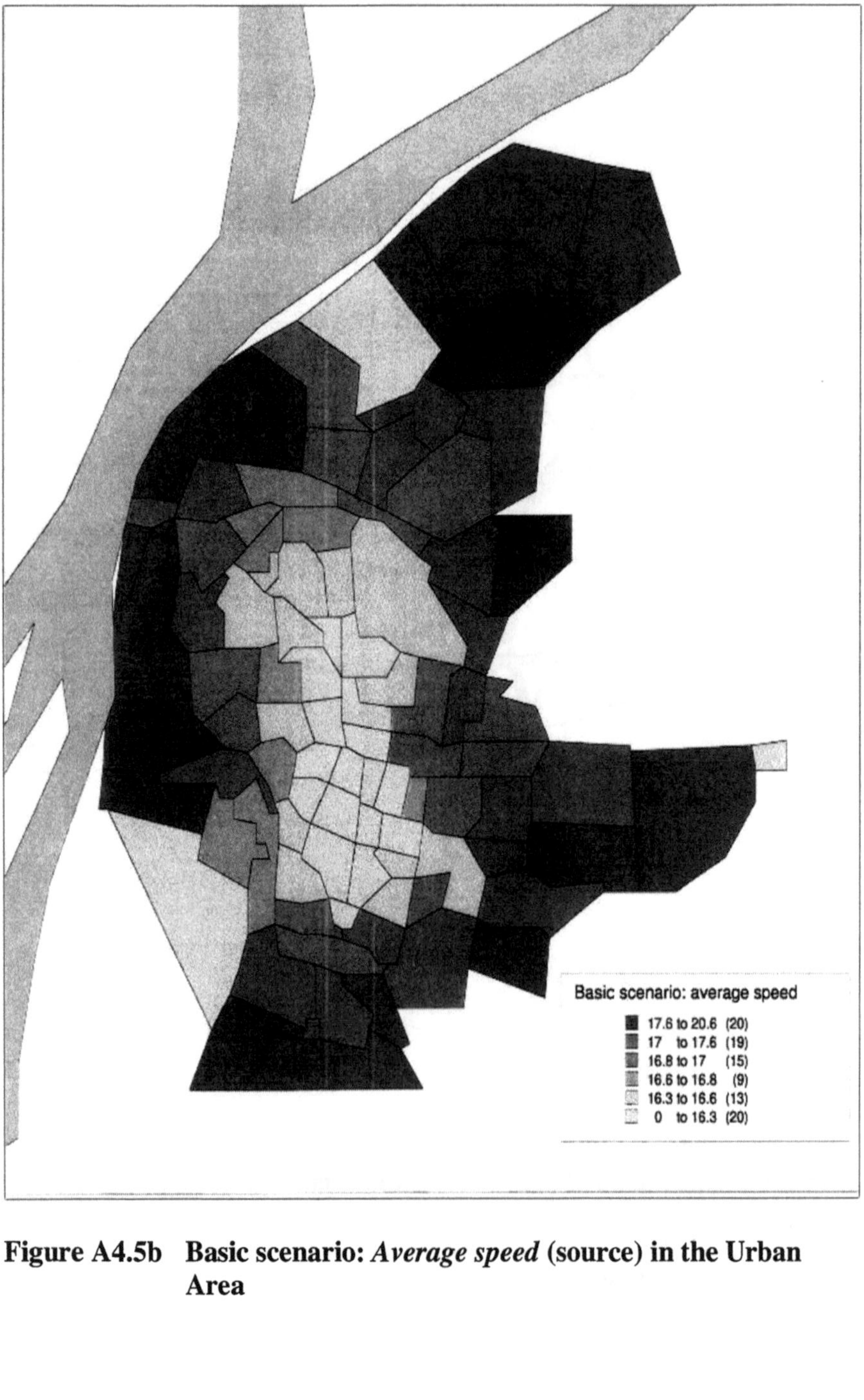

Figure A4.5b Basic scenario: *Average speed* (source) in the Urban Area

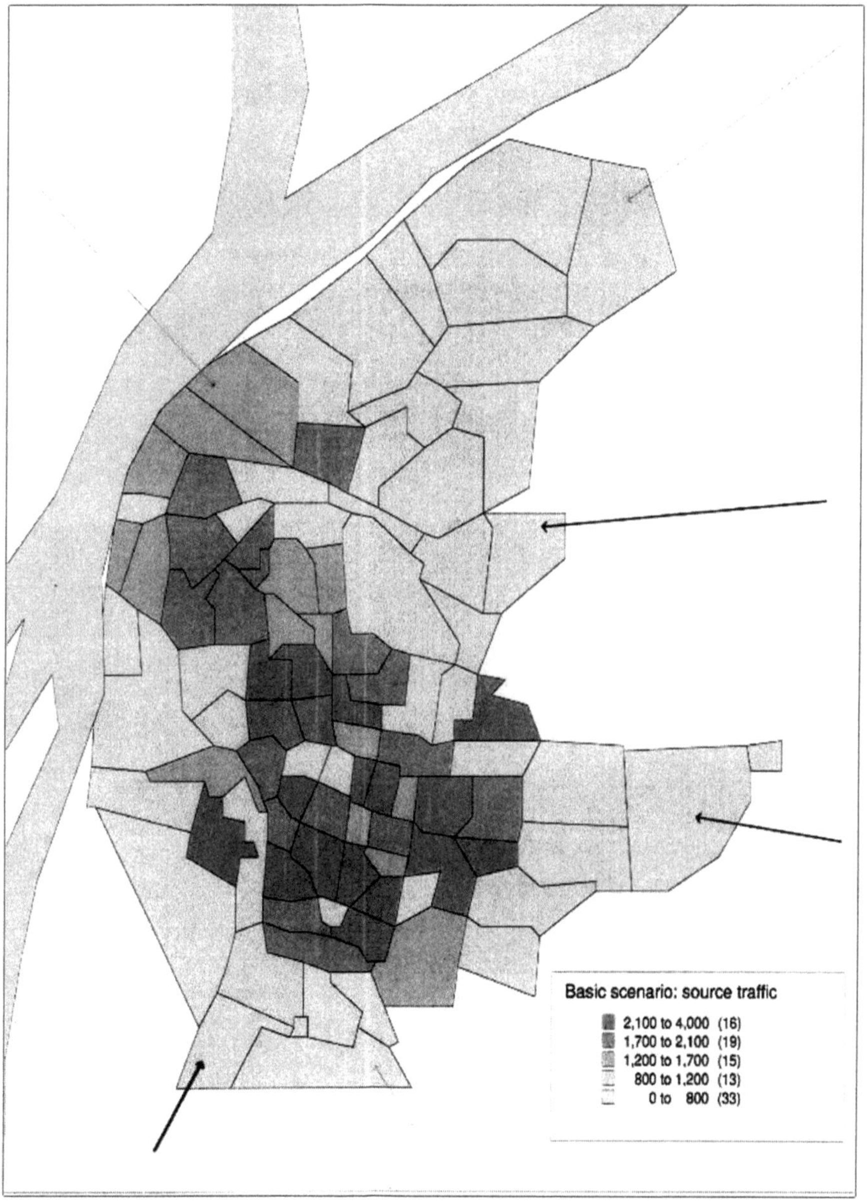

Figure A4.5c Basic scenario: *Source traffic* in the Urban Area

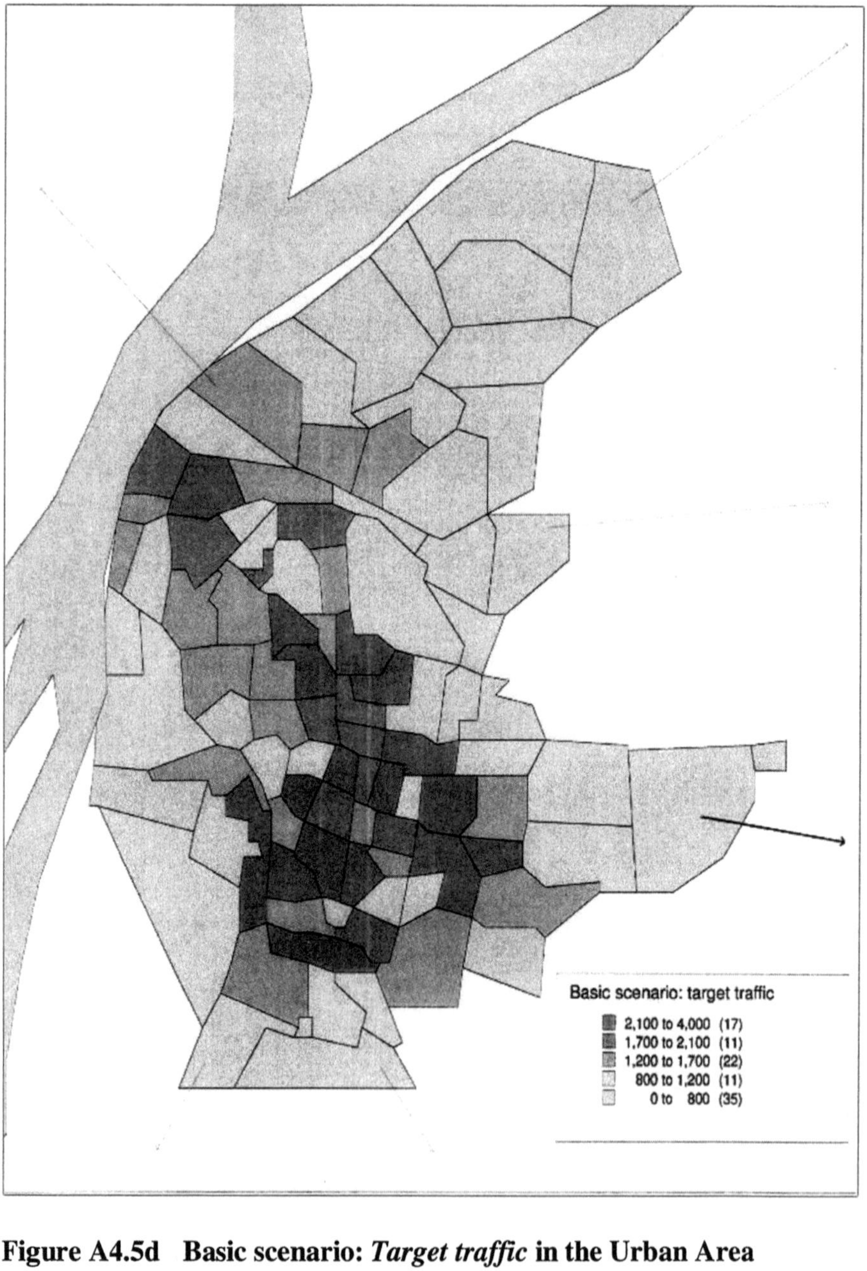

Figure A4.5d Basic scenario: *Target traffic* in the Urban Area

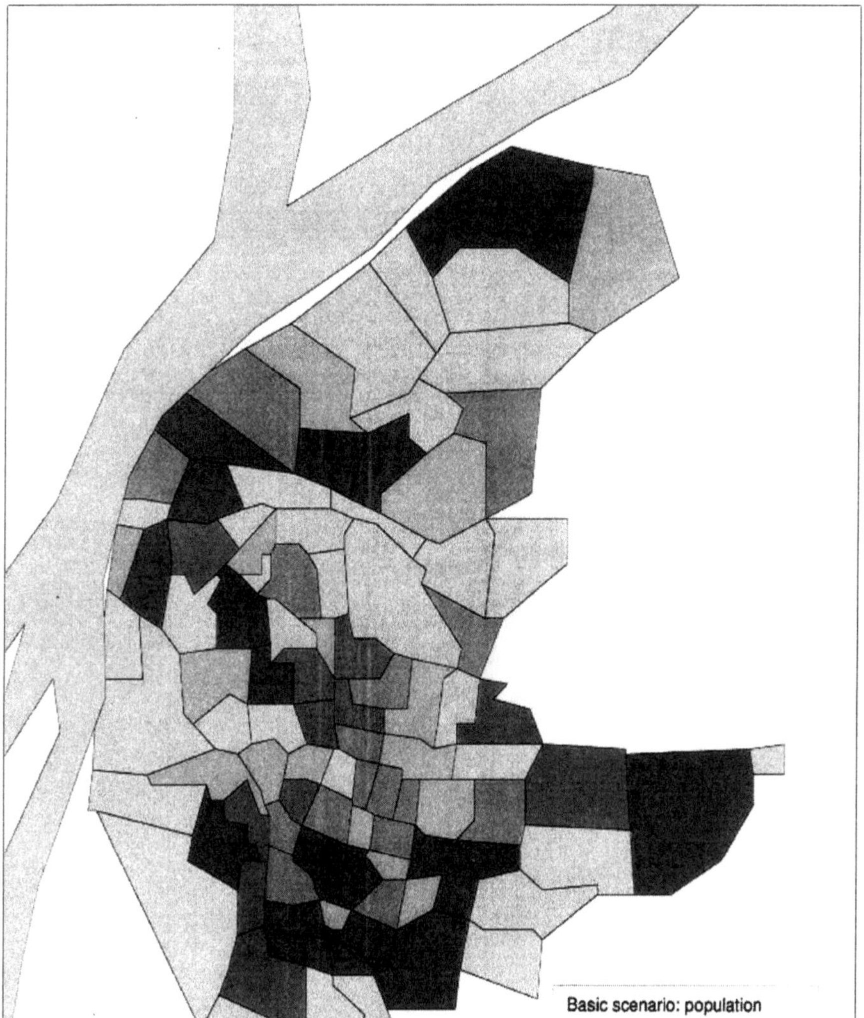

Figure A4.5e Basic scenario: *Population numbers* in the Urban Area

0 to 500 trips | 501 to 1000 trips
1001 to 2000 trips | 2001 to 11210 trips

Figure A4.6a Scenario A: *Assignment* of the street network in the Urban Area

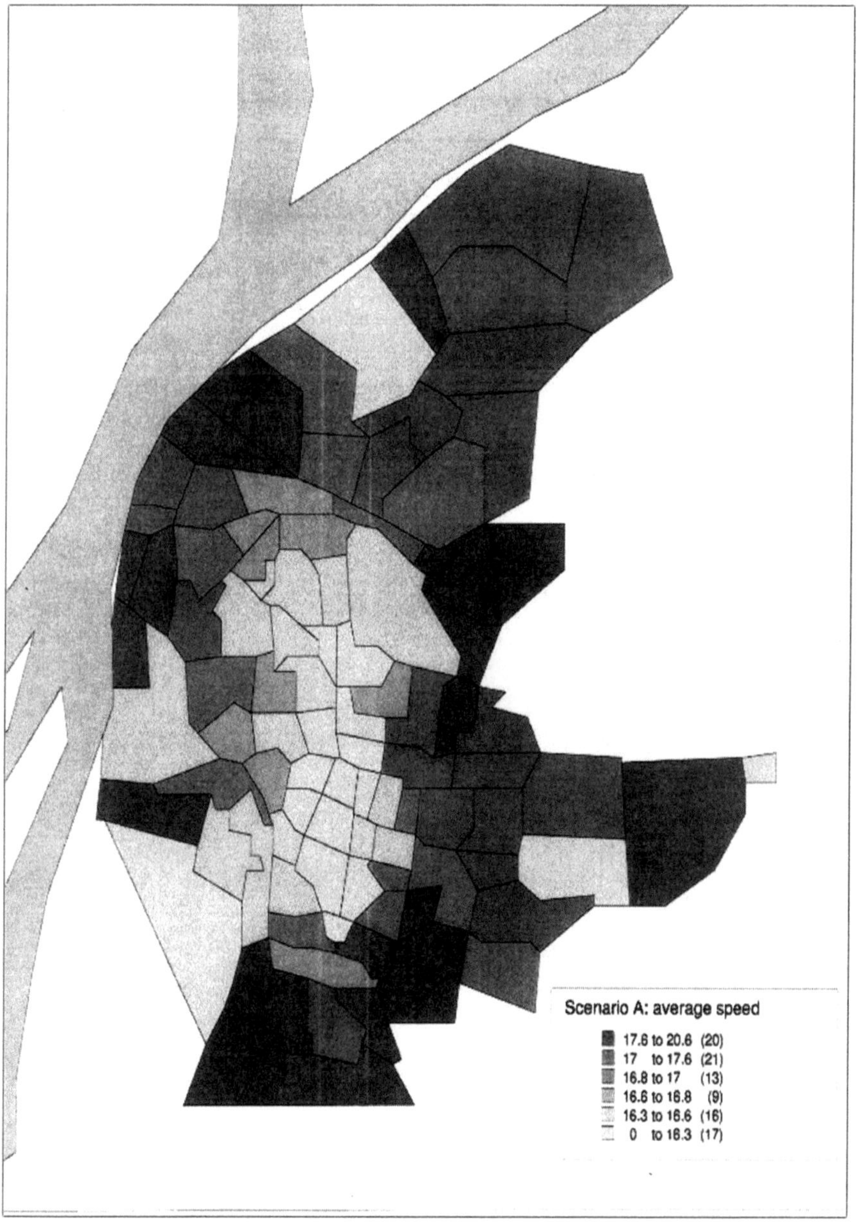

Figure A4.6b Scenario A: ***Average speed*** **(source) in the Urban Area**

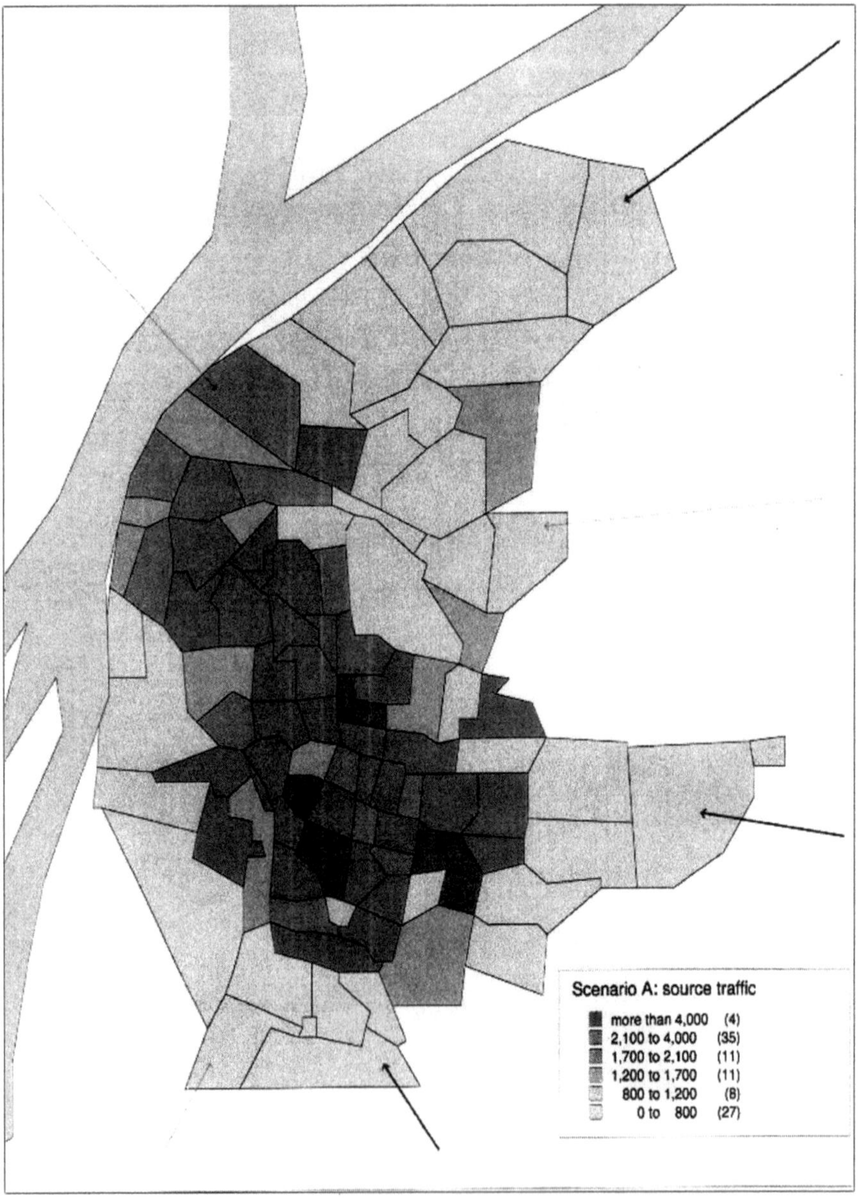

Figure A4.6c Scenario A: *Source traffic* in the Urban Area

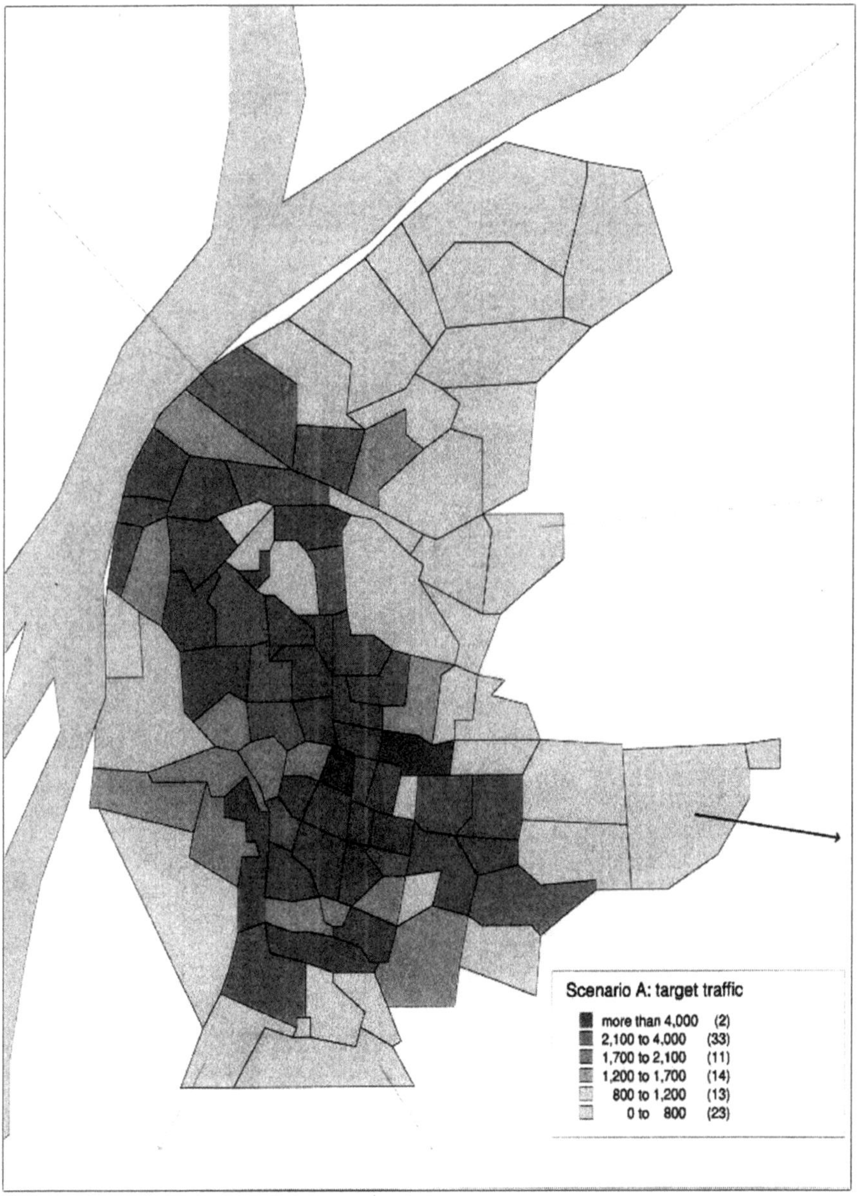

Figure A4.6d Scenario A: *Target traffic* in the Urban Area

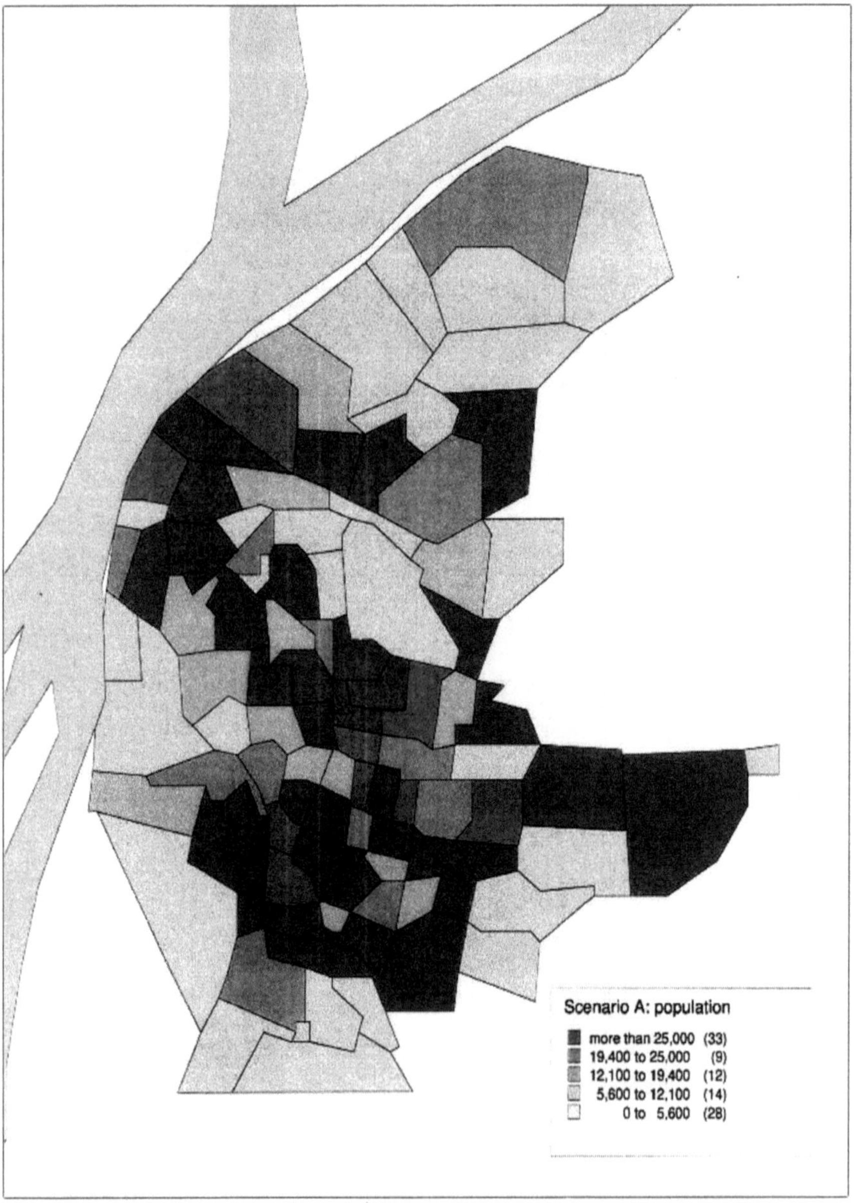

Figure A4.6e Scenario A: *Population numbers* in the Urban Area

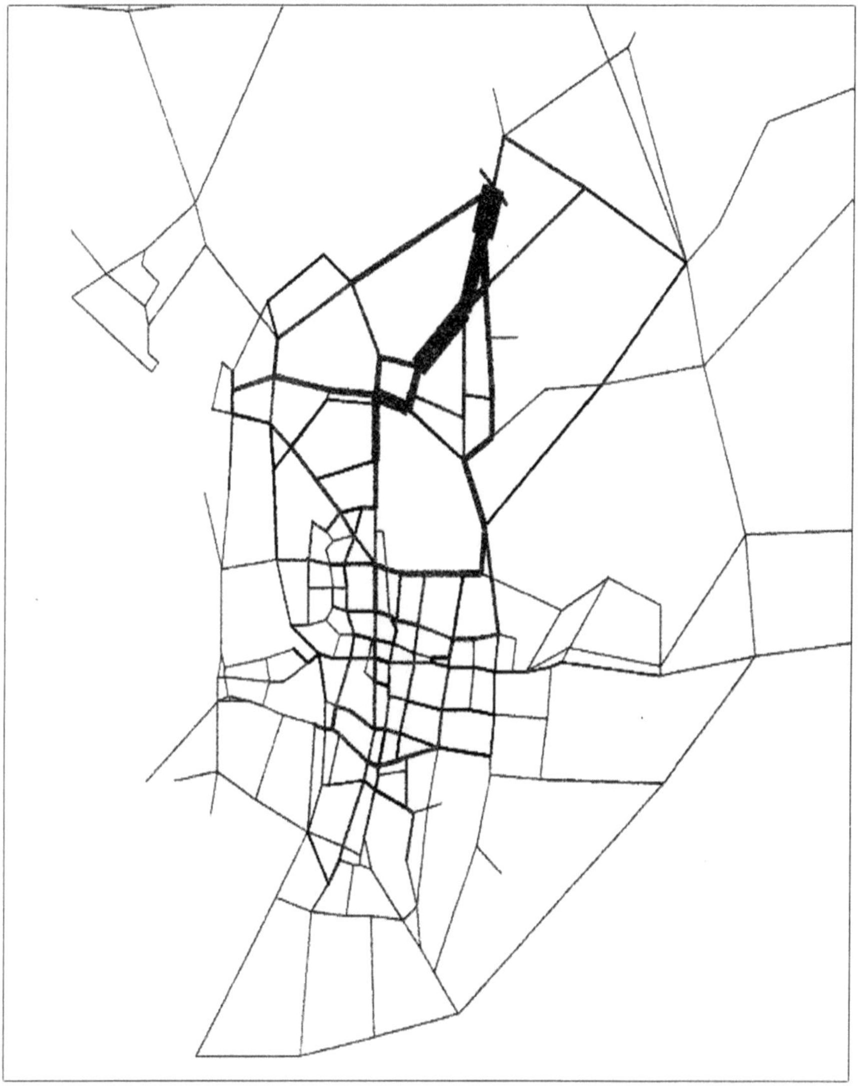

0 to 500 trips
501 to 1000 trips
1001 to 2000 trips
2001 to 9740 trips

Figure A4.7a Scenario B: ***Assignment*** **of the street network in the Urban Area**

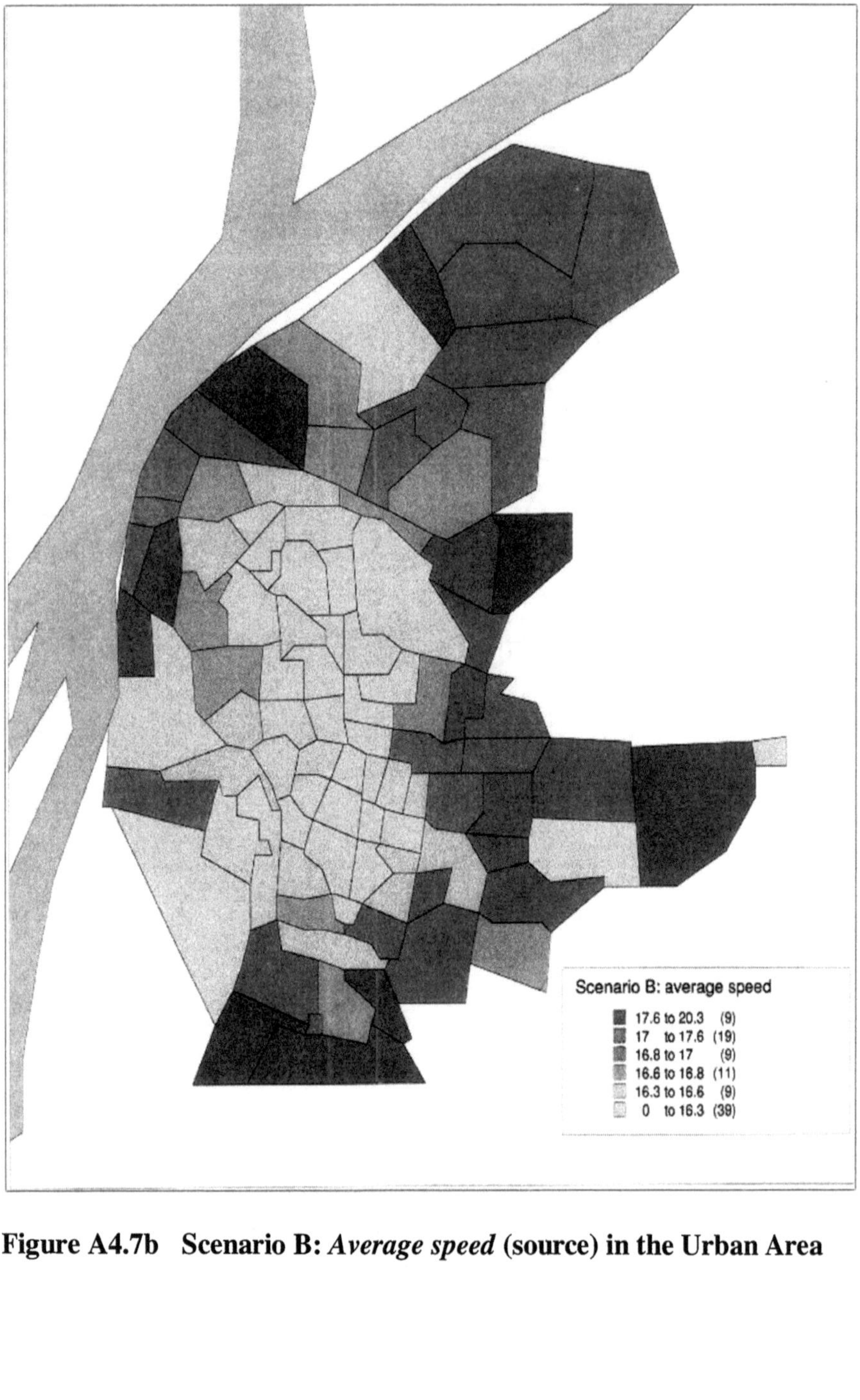

Figure A4.7b Scenario B: ***Average speed*** **(source) in the Urban Area**

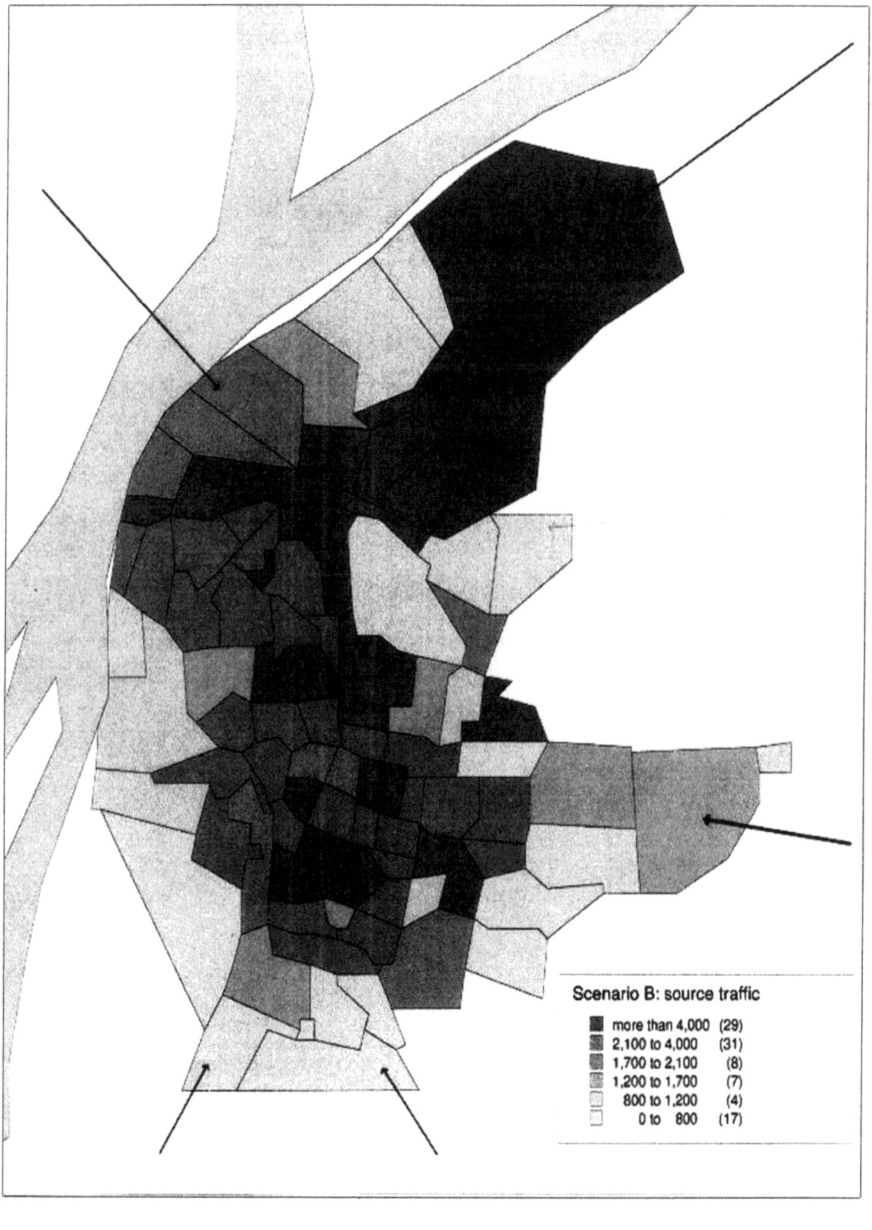

Figure A4.7c Scenario B: *Source traffic* in the Urban Area

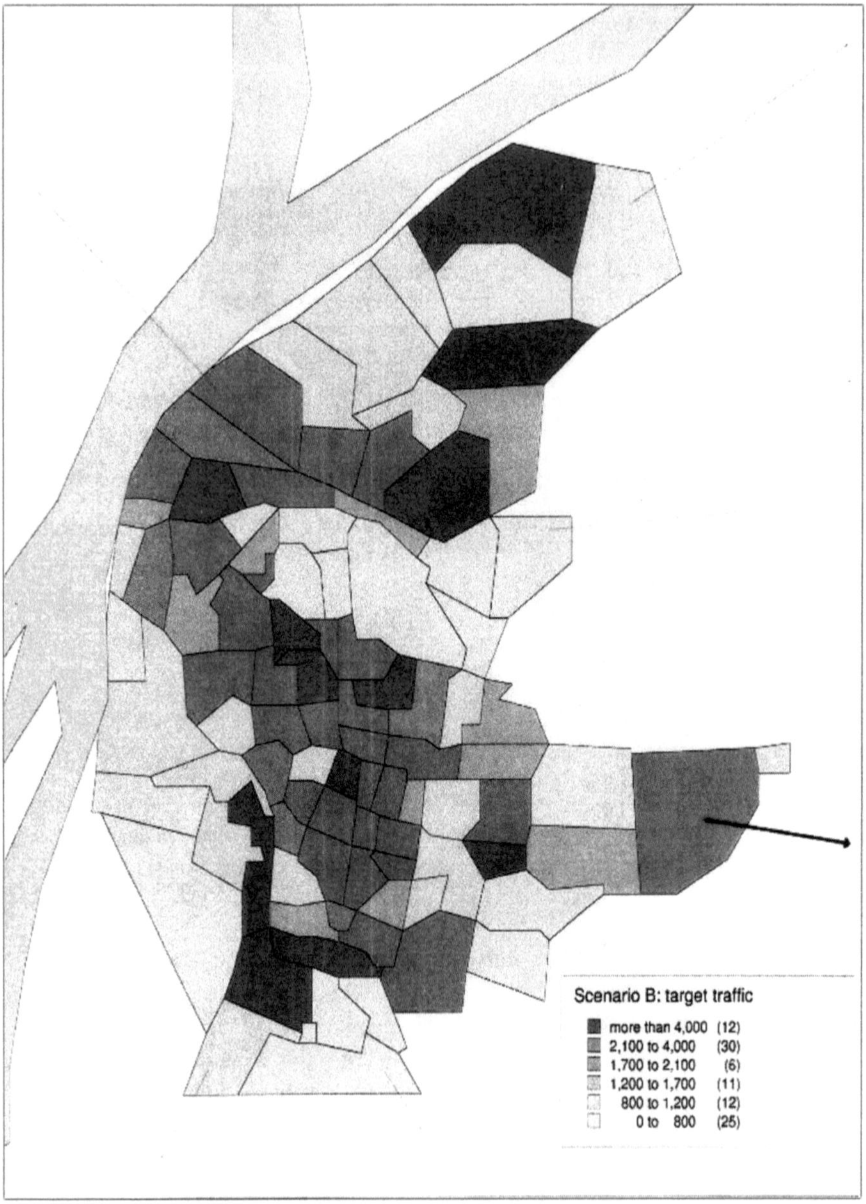

Figure A4.7d Scenario B: *Target traffic* in the Urban Area

Figure A4.7e Scenario B: *Population numbers* in the Urban Area

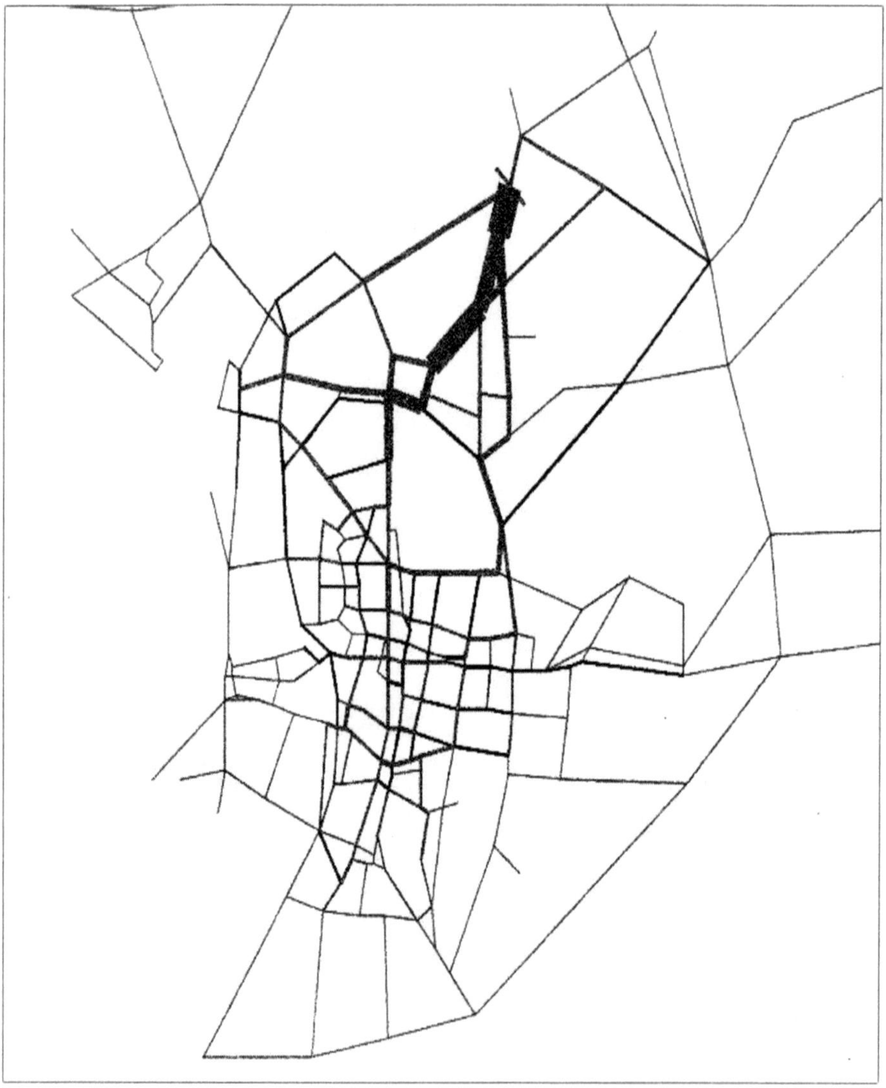

0 to 500 trips
501 to 1000 trips
1001 to 2000 trips
2001 to 54890 trips

Figure A4.8a Scenario C: *Assignment* of the street network in the Urban Area

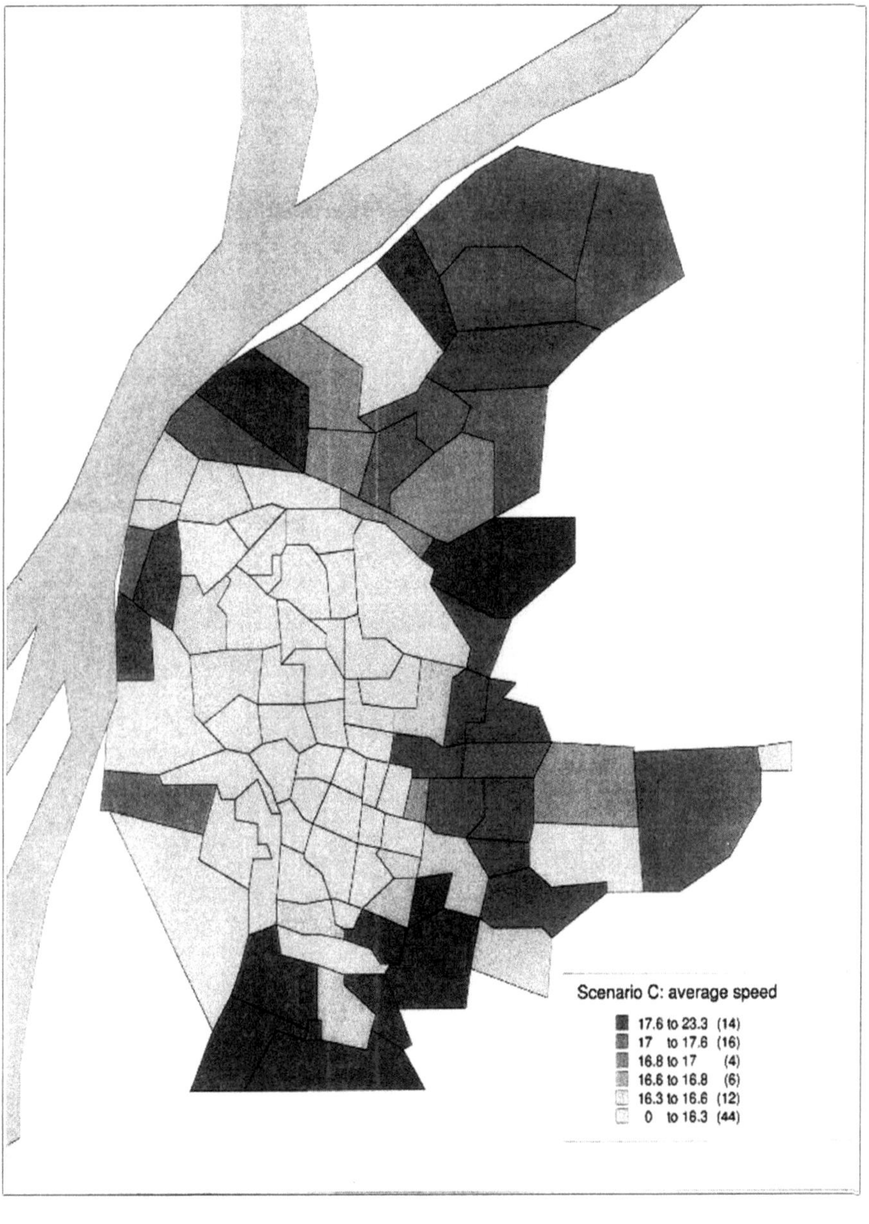

Figure A4.8b Scenario C: *Average speed* (source) in the Urban Area

Figure A4.8c Scenario C: *Source traffic* in the Urban Area

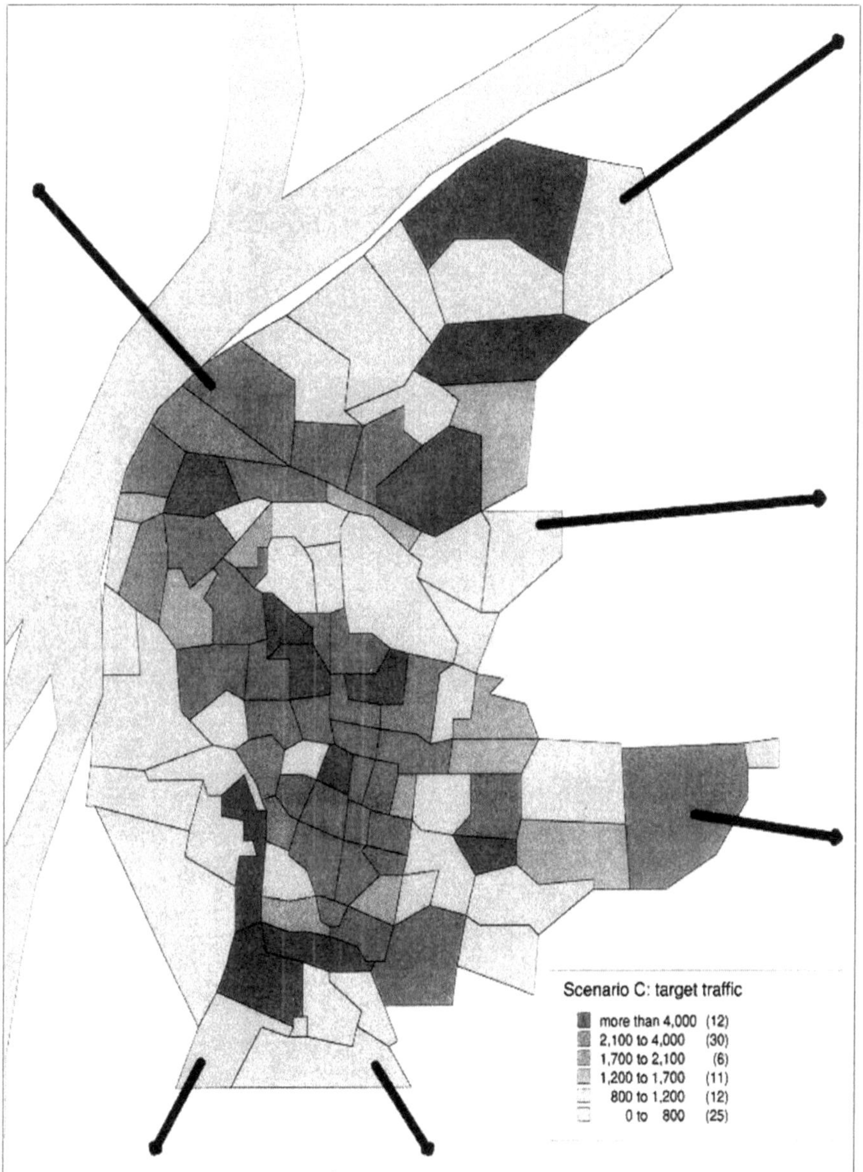

Figure A4.8d Scenario C: *Target traffic* in the Urban Area

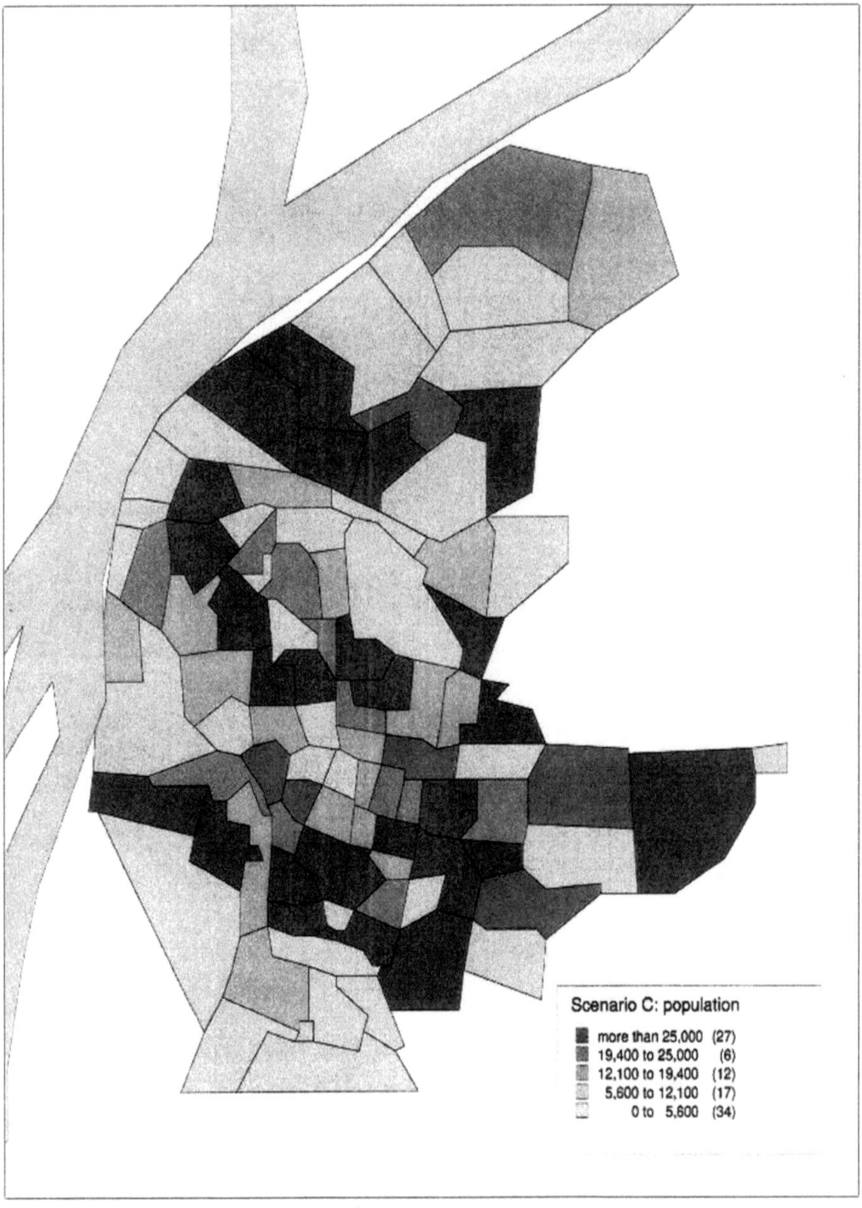

Figure A4.8e Scenario C: *Population numbers* in the Urban Area

References

BUNDESMINISTER VERKEHR 1993: Der Bundesminister für Verkehr (Hrsg.): *Gesamtwirtschaftliche Bewertung von Verkehrswegeinvestitionen*, Schriftenreihe Heft 72, 1993

BUNDESMINISTER VERKEHR 1987: Der Bundesminister für Verkehr (Hrsg.): *Bewertung von Verkehrswegeinvestitionen im ÖPNV*, 1987

BUNDESMINISTER VERKEHR 1996: Qualifizierung, Quantifizierung und Evaluierung wegebauinduzierter Beförderungsprozesse, FE-Nr. 90436/95

DAIMLER-BENZ AG 1990: *Straßenverkehr in China*; Forschungsprojekt 1990

FAZIO/ZANNA 1981: Fazio, R.H.; Zanna, M.P.: *Direct Experience and Attitude-Behaviour Consistency*, in: Berkovitz, L. (Hrsg.): *Advances in Experimental Social Psychology*, **51**, 161-202, 1981

FECHNER 1877: Fechner, G.T.: *In Sachen der Psychophysik*, Leipzig, 1877

FISHBEIN/AJZEN 1975: Fishbein, M.; Ajzen, I.: *Belief, Attitude, Intention and Behaviour: An Introduction to Theory and Research*, Reading, 1975

GAUDRY/MANDEL/ROTHENGATTER 1994: Gaudry, Marc; Mandel, Benedict; Rothengatter, Werner: *Introducing Spatial Competition through an Autoregressive Contiguous Distribution (AR-C-D), Process in Intercity Generation-Distribution Models within a Quasi-Direct Format (QDF)*, paper presented at the Transport Econometrics Conference of the AEA in Calais, 1994

GOODWIN 1994: Goodwin, P.B.: *Traffic Growth and the Dynamics of Sustainable Transport Policies*, ESRC Transport Studies unit, University of Oxford, 1994

GOODWIN 1995: Goodwin, P.B.: *Empirical Evidence on Induced Traffic: A Review and Synthesis*, ESRC Transport Studies unit, University of Oxford, 1995

HAAG 1989: Haag, Günter: *Dynamic Decision Theory: Applications to Urban and Regional Topics*, Dordrecht, 1989

HAAG 1990: Haag, Günter: Transport: A Master Equation Approach, In: Bertuglia, C.S.; Leonardi, G.; Wilson, A.G.: *Urban Dynamics, Designing an Integrated Model*, London und New York, 1990

HAUTZINGER 1982: Hautzinger, H.: *Reiseweite- und Reisezeiteffekte von Geschwindigkeitszuwächsen im Personenverkehr* , in: Internationales Verkehrswesen, **34**, 227-249, 1982

KUTTER 1972: Kutter, Eckhard: *Demographische Determinanten städtischen Personenverkehrs*, Braunschweig, 1972

MEIER 1990: Meier, Eugen: *Neuverkehr infolge Ausbau und Veränderung des Verkehrssystems*, Schriftenreihe des Instituts für Verkehrsplanung, Transporttechnik Straßen- und Eisenbahnbau der ETH Zürich, **81**, 1990

NIJKAMP/REGGIANI 1992: Nijkamp, Peter/Reggiani, Aura: Interaction, Evolution and Chaos in Space. Berlin, Heidelberg, New York, 1992

SACTRA 1994a: The Standing Advisory Committee on Trunk Road Assessment: *Trunk Roads and the Generation of Traffic*, the Department of Traffic, London, 1994

SACTRA 1994b: The Standing Advisory Committee on Trunk Road Assessment: *Trunk Roads and the Generation of Traffic*, Response by the Department of Traffic to the Standing Advisory Committee on Trunk Road Assessment, the Department of Traffic, London, 1994

STEIERWALD/SCHÖNHARTING 1993a: Steierwald, Schönharting und Partner GmbH: *Nachbarschaftsverband Stuttgart, Personen-/Kfz-Verkehr 1990*, Stuttgart 1993

WEIDLICH/HAAG 1983: Weidlich, Wolfgang; Haag, Günter: *Concepts and Models of a Quantitative Sociology*, Berlin, Heidelberg, New York, 1983

List of Senior Advisors, Contributors and Cooperating Partners

Senior-Advisors

Prof. Dr. Dr. h.c. mult. I. Prigogine
Instituts Internationaux de Physique et de Chimie
Campus Plaine ULB, CP. 231
Bd. Du Triomphe
1050 Bruxelles, Belgium

Prof. Dr. Dr. h.c. mult. H. Haken
I. Institut für Theoretische Physik
Universität Stuttgart
Pfaffenwaldring 57/IV
70550 Stuttgart

Contributors

Prof. Dr. F. Englmann
Institut für Sozialforschung
Universität Stuttgart
Keplerstr. 17
70049 Stuttgart

Dipl. Phys. K. Grützmann
Steinbeis-Transferzentrum Angewandte Systemanalyse
Rotwiesenstr. 22
70599 Stuttgart

Prof Dr. G. Haag
Steinbeis-Transferzentrum Angewandte Systemanalyse
Rotwiesenstr. 22
70599 Stuttgart

Prof. Dr. P. Nijkamp
Dept. of Economics
Free University Amsterdam
De Boelelaan 1105
1081, HV Amsterdam, The Netherlands

Prof. Dr. Y. S. Popkov
Institute for System Analysis
Academy of Sciences of Russia
9, Prospekt 60-let Oktyabria
117312 Moscow, Russia

Prof. Dr. A. Reggiani
Dipartimento di Scienze Economiche
Università degli Studi di Bologna
Piazza Scaravilli, 2
40126 Bologna, Italy

Dr. rer. nat. T. Sigg
II. Institut für Theoretische Physik
Universität Stuttgart
Pfaffenwaldring 57/III
70550 Stuttgart

Prof. Dr. Dr. h.c. W. Weidlich
II. Institut für Theoretische Physik
Universität Stuttgart
Pfaffenwaldring 57/III
70550 Stuttgart

Cooperating Partners

Prof. Dr. Deng Wei
Southeast University
Institute of Traffic and Transport
Sipailou 2
210096 Nanjing, VR China

Prof. Dr. Wang Wei
Southeast University
Institute of Traffic and Transport
Sipailou 2
210096 Nanjing, VR China

MIX
Papier aus verantwortungsvollen Quellen
Paper from responsible sources
FSC® C105338

If you have any concerns about our products,
you can contact us on
ProductSafety@springernature.com

In case Publisher is established outside the EU,
the EU authorized representative is:
Springer Nature Customer Service Center GmbH
Europaplatz 3, 69115 Heidelberg, Germany

Printed by Libri Plureos GmbH
in Hamburg, Germany